selected topics in biology

TOOLS AND TECHNIQUES

G. H. Harper

Nelson

Thomas Nelson and Sons Ltd
Nelson House Mayfield Road Walton-on-Thames Surrey KT12 5PL

51 York Place Edinburgh EH1 3JD

Yi Xiu Factory Building
Unit 05–06 5th Floor 65 Sims Avenue Singapore 1438

Thomas Nelson (Hong Kong) Ltd
Toppan Building 10/F 22A Westlands Road Quarry Bay Hong Kong

Thomas Nelson (Kenya) Ltd
P.O. Box 18123
Nairobi Kenya

The cover illustration is a scanning electronmicrograph of the face of a male, yellow fever mosquito (*Aedes aegypti*). (Courtesy of Dr Tony Brain/Science Photo Library)

First published by Thomas Nelson and Sons Ltd 1984
Reprinted 1984

ISBN 0-17-448092-X
NCN 21-KSC-3308-02

Diagrams by Illustrated Arts Ltd, Sutton, Surrey
Printed in Hong Kong

General editor's preface

The books in this series are written specifically for A-level students who wish to pursue certain areas of biology in depth. We hope the books will be particularly useful to those who are taking special papers or entrance examinations to Oxford or Cambridge.

The writers of books of this kind face a dilemma. On the one hand they want to provide plenty of material to challenge the keen sixth former. On the other hand they do not want to smother the poor student with a plethora of detail which he or she will shortly meet again at university. The aim of this series is to broaden the student's view of biology without trespassing on university territory. The emphasis is on providing a wide range of interesting examples and studies which we hope will stimulate thought and enquiry. At the same time we hope that the books will prove enjoyable as well as informative.

In preparing the series we have been highly selective in our choice of topics. We have confined ourselves to those topics which are difficult to cover adequately in a basic textbook and for which we consider there is a need for appropriate books at the sixth form level.

M. B. V. Roberts

Author's preface

An important book appeared in 1965: J. A. Ramsay's *Experimental Basis of Modern Biology*. It was designed for sixth formers preparing for Cambridge and Oxford scholarship examinations.

Ramsay's book resulted from a feeling of frustration. His candidates could rattle off the details of metabolic pathways, but when asked where this knowledge came from they were usually clueless. In Ramsay's opinion this meant there was something wrong. Training a biologist did not mean just teaching the conclusions which researchers had arrived at. To be any good the training should also emphasise the role of observation and experiment, the importance of errors in collecting facts, and how to judge theories and evidence.

These opinions have been constantly in my mind while writing this book. Unlike Ramsay's book, however, this one is aimed at all A-level students in biology. While it might appeal especially to those of you who hope to become biologists, other students should gain much from it. For one thing, the examples in the book may increase the breadth of your knowledge, or help you to understand topics that have seemed difficult. In addition, many of the A-level examination boards now list a variety of instruments and methods in their syllabuses. Most or all of these are described in this book.

The title *Tools and Techniques* might suggest this is a practical handbook, telling you how to use the various instruments and methods. For that kind of information, however, you really need to look at the manufacturers' instructions that come with the instruments. In this book the aim has been rather to explain the basic principle behind each method. You are not told, for instance, how to set up an oscilloscope, but you *will* learn how it works and why it is so valuable.

I have enjoyed writing this book, and I hope you enjoy reading it. You could go through it from beginning to end if you wish. On the other hand, most of the chapters are self-contained, so that they can be taken in any order. At a later stage you may feel like going on to Ramsay's book or to others listed in the References at the end.

G. H. Harper

Contents

Acknowledgements

In the preparation of this book, various chapters were read by Dr Geoffrey Arnold, Dr Roy Harden Jones, Andrew Lloyd, Peter Mawby and Dr Jaleel Miyan. In addition, Michael Davison, John James and Dr Brenda Weakley showed me their electron-microscope facilities at Dundee University. Some useful references were supplied by Richard Steele. Hetty Harper assisted with proof reading, and must be the only person who will ever read the whole book out loud. For all this help I am most grateful.

The guidance and encouragement of Donna Evans, Elizabeth Johnston and David Swinden at Thomas Nelson are much appreciated; and special thanks are due to the editor, Michael Roberts, for his patience and expertise.

All opinions expressed in this book are my sole responsibility.

Acknowledgement is due to the following for permission to reproduce photographs.

Fig. 46, p. 93, Dr Brenda Weakley, Department of Anatomy, Dundee University; and Churchill Livingstone.

Fig. 47b, p. 95, Michael Davison, Department of Biochemistry, Dundee University.

Fig. 47d, p. 95, The Electron Microscope Unit, Zoology Department, Edinburgh University.

Fig. 49, p. 98, Dr J. B. Finean, Department of Biochemistry, Birmingham University; and *School Science Review*.

Fig. 50, p. 100, Dr A. M. Glauert, Strangeways Research Laboratory, Cambridge; micrographs taken by Prof. L. D. Peachey, Department of Biology, Pennsylvania University; and *Journal of Microscopy*.

Fig. 51a, p. 102, John James, Department of Anatomy, Dundee University.

1 Biology as a science

Observations are the foundation of biology. This book is all about how to make them.

Two well-known features of any science are the 'facts' and 'theories'. Facts are the immediate results of scientific observation, and theories are statements which generalise or explain groups of facts. Any science involves both of these, not just one.

For instance, a bird-watcher may go out regularly and collect masses of facts. He or she might list all the birds seen each day. But unless an attempt is made to find patterns in the facts or invent explanations for them, the bird-watching is not proper biology; it is merely observation. On the other hand, a mathematician may sit at a desk and construct elegant theories all day long. These theories are about abstract ideas such as numbers, not real things such as cells and organisms. Therefore, mathematics is essentially theorising, and not a science.

In contrast with bird-watching and mathematics, biology is a science, because it involves both observing facts and devising theories about them.

In the first three chapters, I consider how observations fit into biology, and what is needed to make an observation scientific. The rest of the book describes various ways in which modern biologists collect their facts.

Theories

Just now it was stated that theories generalise or explain facts. Suppose some toadstools are measured. If we say that their heights are in the range 10–12 cm, we are merely **generalising** observations; no reason is given for the facts. But suppose now a reason is suggested: for instance, the toadstools are growing in grassland about 7 cm high, and 10–12 cm is the minimum height necessary for good spore dispersal (figure 1). Now we have an **explanation** as well as a generalisation, and each of these is a theory about the toadstools.

In talking about science, three words often cause confusion: 'hypothesis', 'theory' and 'law'. These are all names for what have so far been called theories. Sometimes people use the world 'law' for generalisations which do not explain anything. An example is Allen's Rule: in a

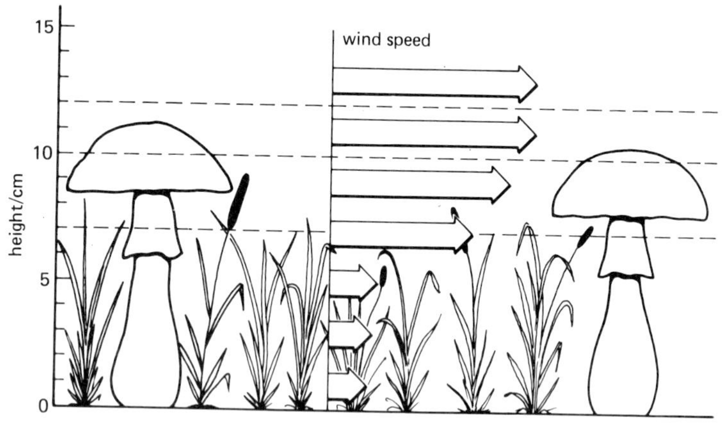

Figure 1 Toadstools 10–12 cm high raise their caps above grassland of average height 7 cm, so taking advantage of the higher wind speeds. Wind speed is indicated by the length of the arrows.

group of related endothermic animals, the sizes of ears, legs and tail are smaller in cooler climates. On the other hand, the word 'theory' may be used for complicated or wide-ranging explanations such as Darwin's theory of natural selection.

Most often a hypothesis is thought to be an explanation or generalisation which is not yet thoroughly tested; it becomes a theory or a law as we become more confident that it is true. The word 'law' is not used much in this book, but 'hypothesis' and 'theory' will be taken to mean much the same as each other. Any generalisation or explanation can be called a hypothesis or a theory.

Did mammals evolve from reptiles or directly from amphibians? Most biologists agree that they evolved from reptiles, and this could be called a hypothesis or theory. It is certainly not an observed fact, because we have no record of it from someone there at the time. But as a theory it is unusual, because it refers to a single event that happened once, if it happened at all. Most theories instead refer to an infinite number of facts. Each theory generalises or explains so many facts that it would be quite impossible even for a whole army of biologists to check the truth of them all. The importance of this needs to be explained.

How much information?

In the toadstool example let us assume that only ten plants were measured. Their heights were respectively 11, 12, 11, 10, 11, 12, 11, 10, 11 and 12 cm, so we conclude that the heights of those ten plants were in the range 10–12 cm. This conclusion is reached by a process known as **deduction**. Deduction is a method of argument using strict rules, and these guarantee that the conclusion is true if the facts themselves are true. It is therefore a powerful tool of thought, and it explains the importance of mathematics, since that subject uses only deduction. The role of deduction in biology will become clear in the next chapter.

As a generalisation, though, the 10–12 cm range of the ten heights is not very important. This is because it contains less information than the ten individual observations. It does not even tell us the height of each plant, or what the average height is.

What would normally happen in practice is that a biologist would measure a sample of the toadstools in a few chosen areas. He would assume that the heights were the same for all the other toadstools of that species over a much larger area. He would probably also assume that the same height range would be found in other years. So this generalisation, based on a measured sample, is meant to apply to a vast number of other individuals. You will appreciate that it contains much more information than the original facts. It says something about an infinite number of organisms.

Does this theory or generalisation result from a deduction? The answer must be 'no'. Deduction produces true conclusions if properly used, but there is a price to be paid for this advantage. The price is that a deduction can never contain more information than was put into the argument in the first place. The deduction about the ten heights only refers to the ten organisms, not to eleven or to 1000.

Therefore, a deduction may be true, but it cannot increase the amount of information. In contrast, a scientific theory does increase the amount of information. But there is again a price to be paid: a theory can never be *proved*. For instance, it is obviously impracticable to measure an infinite number of toadstools, including those that will be in existence at any time in the future.

A hierarchy of theories

So far, observation and two kinds of theory have been mentioned. These can be arranged in a line: observation, generalisation and explanation. But this is only part of a more complicated arrangement which is outlined in figure 2. The five kinds of information form a continuous series, and in chapter 3 it is explained why there is actually no hard and fast division between them.

The **sense data** are the original items of information reaching our eyes, ears and other sense organs. They may consist of patterns of light and

Figure 2 The hierarchy formed by the different kinds of theory and observation.

Kind of information	Examples	
Explanation using unobservable concepts	Theories about genes	
Explanation using only observable concepts		Theories about spore dispersal and the influence of vegetation on wind
Generalisation	'A certain cross always gives a 3 : 1 phenotype ratio'	'Mature sporophytes of this species are always 10–12 cm high'
Observation	'This fruit fly has red eyes'	'This toadstool is 10.5 cm high'
Sense data	Seeing red colour on part of an object	Matching a shape against a mark on a ruler

sound waves which we then interpret to mean that 'this toadstool is 10.5 cm high', 'this fruit fly is male and has red eyes', and so on. It is only when the sense data are interpreted and expressed in language that they become observations which can be communicated to other biologists.

The next stage is usually to summarise observations in a much wider generalisation. Then, when it comes to inventing explanations, these are of two kinds. In the simplest case, as in the toadstools, it may be quite easy to observe all the concepts that the explanation refers to. The height of the grassland can be measured if a suitable technique is chosen; so can the wind speed at various heights above the ground, and the distance spores are carried by the wind.

However, not many important theories in biology use only observable concepts. Mendelian genetics is all about genes, but genes cannot be observed. Nor can we directly observe such things as atoms, evolution, energy, trophic levels, niches and homeostasis. But it is important to make them observable if possible, so that the close relation between 'fact' and 'theory' in biology can be maintained. This difficulty is discussed in chapter 3 (see p. 26).

Where do theories come from?

It used to be thought that there were methods for producing theories. Observations had to be collected first, and then a set of rules was used to draw out the correct theory from the facts. This process was called

induction. However, few people now accept this view, if only because it is not the way scientists work.

Most books describing the methods of science do not say how theories are produced. One might think that more could be learnt from the scientists who devise theories. The most obvious sources to turn to are the scientific papers in which the theories are first announced. Unfortunately, these papers give little idea of how investigations are carried out. There is a famous book on science called *The Logic of Scientific Discovery*, but despite its title the whole point of the book is that there is *no* 'logic' of discovery. In short, there are no rules for constructing theories, and though perhaps surprising this opinion is now very widely held.

Theories begin as guesses. Many of them arrive in the minds of biologists in a quite spectacular manner. It is well known how both Alfred Russel Wallace and Charles Darwin, independently of each other and at different times, suddenly realised how natural selection could explain evolution. The idea was suggested to both by their reading of Malthus's book on political economy, and this may mean that the theory of natural selection came from outside biology. Theories often do include an analogy between a biological phenomenon and something quite different. But the important point is that the new idea seems to arrive in the mind, often suddenly and unexpectedly, from nowhere in particular. It is an inspiration, an imaginative leap to a new understanding of the biological facts. Careful calculation and methodical argument are not the qualities of mind needed to create theories; it is more useful to have a fertile imagination which can conjure up novel ideas from nowhere. No wonder there are no rules for producing theories. This also explains why careful observations do not need to be made first.

This is not to say that creators of theories do not have ways of encouraging their creativity. W.I.B. Beveridge's readable book *The Art of Scientific Investigation* has a section called 'Technique of seeking and capturing intuitions'. This includes several useful recommendations. Novel ideas often come, for instance, during comparative inactivity: examples are shaving, travelling, bathing, or lying in bed. Indeed, it has been plausibly suggested that this explains the famous anecdote about Archimedes. You will remember the story that he thought of 'Archimedes' Principle' while in the bath, and immediately lept out and ran naked down the street shouting 'Eureka!' Most people have assumed that the theory came to him in the bath because it concerns floating bodies, but this may have been only a coincidence. The real significance of the bath would perhaps have been the effect of relaxation on his creative imagination.

Inventions and discoveries

It might seem that there could not be much confusion between inventing theories and discovering new biological facts. For example, Darwin discovered the finches on the Galapagos islands, and he also invented a theory of heredity called pangenesis which nobody now takes seriously. This is clear enough. But many people would accept not only that Watson and Crick invented the theory of DNA's structure, but that they also actually *discovered* the structure. In other words, when a theory becomes so widely accepted that no one doubts it any longer, the word 'discovery' tends to be used. However, this should not mislead us into thinking the theory no longer needs to be questioned. It would be better if 'discovery' were only used for objectively observed facts.

Problems

Before a biologist thinks up a new theory, he has usually become interested in the particular area of study. The theory comes to him only after he has read other people's work and has thought a good deal about the subject. Various things can stimulate interest in a subject, and usually these take the form of problems.

Problems may sometimes be practical needs, especially in applied biology. No one doubts that cures for cancer ought to be found if at all possible. Similarly, any other disease for which a cure or prevention is needed represents a problem for some biologist to solve.

Other obvious problems existed in the 1950s in genetics. The behaviour of genes was well-known, but the form which the genetic information took was unknown. It was generally thought to be in the form of a complex molecule – protein or nucleic acid. The problems were therefore to identify which chemical it was, and then to find out how it carried the information. Watson and Crick were spectacularly successful in solving these problems. But skill was required not just for solving the problems; it was also needed in selecting the problem to work on in the first place. It is almost impossible to give advice on how to select a problem likely to lead to a satisfactory solution. Mostly, the biologist just follows his own interest and has an intuitive knack of picking on questions waiting to be answered.

Problems fall into three groups. One is based on practical need, such as curing diseases or increasing food production. A second arises from patterns that have been found in some facts; the problem is to find an explanation. The generalisations in figure 2 are examples of such patterns. The third kind of problem is rather different, for it appears when there is a deviation from a pattern. For instance, there is a simple relation between the size of different mammals' brains and the mass of their bodies; but a human brain is much larger than would be expected for a mammal of human mass. So, why is this?

A special case of deviation from a pattern will crop up again in the next chapter. It happens when a theory is being tested. A theory may lead us to expect a certain set of results from an experiment, and indeed most of the results may be satisfactory. But if one or two are significantly different from the expected pattern, a problem has immediately been created, or rather two problems: do we now have to abandon the theory, and how is the deviation from the expected pattern to be explained?

These deviations from expected results are called **anomalies**, and the investigation of anomalies has led to some of the most important advances in biology. For instance, it was found that some plants could be infected with a certain disease; this was done by transferring sap from a diseased plant to a previously uninfected one. Normally it is possible to isolate the bacterium or other microbe from infected sap by filtering, but in this case not even the finest filters could remove the infecting organism. This unexpected result was an anomaly. Further investigation of it led to the discovery of viruses.

Biological knowledge

Biology aims to produce a true account of how living matter and organisms work. The account is intended to be as detailed and complete as possible, but there are reasons for thinking that the account will never be complete.

Science works only because certain rules are obeyed, for example rules concerning objectivity of observation (see p. 25). It is because of these rules that biological knowledge has to take a certain form. This can be illustrated first by considering those rules of observation.

If an observation is objective, another biologist should be able to make the same observation if the same procedure is used. However, some things we are all aware of simply cannot be observed objectively. These include a person's feelings, thoughts or intentions. The best that another observer, such as a psychologist, can do is to record the person's behaviour and speech. In case this is not clear, ask yourself whether an animal you are watching has feelings, thoughts and consciousness. You cannot observe, still less *measure*, any of these. Indeed, a clever engineer could probably build a machine which imitates exactly all the actions seen in the real animal.

So, there are certain properties of living things which simply do not enter the sphere of investigative biology. They must be ignored in the biological account of life. Even in the things that biology can tackle, there is a restriction on how they are described. You will remember that, apart from theories about historical or prehistoric events, biological theories refer to large sets of things, so numerous that they cannot all be observed. As a result nothing truly individual or unique is recognised in biological theories. Individual persons are given individual names, such as 'Zaphod Beeblebrox', but physiologists do not publish detailed studies of an individual mitochondrion called Sally or no. 2358. In behaviour studies, animals may be given names and studied as individuals, but the important observations are those which in the end are true for a whole set of animals. The individual differences are ignored.

Biology therefore gives only a partial and simple view of the biological world. Whole phenomena such as subjective experiences are missed out, while those things that are included must be identified as members of sets. These features of biological knowledge give it a distinctive form, which can be contrasted with that of a biography. Biology and biography are both ways of describing facts; but unlike biology, a biography does deal with the subjective experiences of individuals. On the other hand, biological knowledge depends on the methods of science, no matter what biologists actually believe the natural world is like. In practice, methods and beliefs are closely entangled, and the latter need to be mentioned.

Beliefs

The confusion between methods and beliefs can be illustrated by animal behaviour. In the school of thought called behaviourism, theories are not made about animals' subjective experiences such as thoughts and feelings. However, there are two forms of behaviourism. In one, the behaviourist says that thoughts and feelings in animals should be ignored because they cannot be observed. At the same time, the animal may or may not have these experiences; we simply do not know. The other kind of behaviourist uses the same method, but he has a definite belief. He is quite sure that animals do not have thoughts and feelings, so that it is meaningless to talk about them.

Here it is quite clear that the method and the corresponding belief are distinct. Figure 3 summarises five examples like this. The five methods listed are used by all biologists, but the beliefs are optional. A biologist could be a good scientist by following the methods even if he did not hold any of those beliefs. He might believe that animals do have thoughts and feelings; that individual spinach chloroplasts are as different from each other as the people in a crowd; that nature is immensely complicated even at the most basic level; that some phenomena do not have causes; and that others are indeed caused by supernatural acts of God.

Figure 3 Some of the methods used in biology, with the corresponding beliefs.

Method	Belief
Animals' subjective experiences are not allowed in theories	Animals do not have subjective experiences
Objects must be identified as members of sets or classes	Objects like atoms do not have individual differences
Occam's Razor: simple theories are preferable to complicated ones	Nature is fundamentally simple and well-ordered
One should try to find the causes of of things	Universal causation: everything has a cause
Supernatural causes are not allowed in theories	There are no supernatural causes

However, it seems that most biologists do share most of the beliefs listed in figure 3. To check this you might like to survey opinion among the science students and staff in your school. It is these shared beliefs about how the natural world works that most obviously separate scientists from other people. This is well-illustrated by the long-running debate about the origin of species. The main competitors are the evolutionists and the special creationists. They all agree with origin theory, which states that species did come into existence, but the evolutionists are scientific because they believe that no supernatural causes were involved. On the other hand, the special creationists are not scientific because they are sure that God created the species using some means other than natural processes. It is these beliefs which make one lot scientists and the others not.

The beliefs in figure 3 are something like scientific theories. After all, they describe how the natural world works. But they are not really scientific theories, because there is no way in which they can be tested. In fact, it is quite impossible to disprove any of them. For instance, universal causation can never be disproved, because, even if something seems to have no cause, it is always possible that there is a cause which we have not found.

So, each person must make up his own mind about these beliefs. His opinions probably determine whether he wants to become a scientist. They are important because they influence the way biologists think, and the way in which the scientific view of the biological world is built up. It is commonly thought among scientists that the scientific view will eventually provide a true description of everything worth knowing: but to accept this requires a considerable act of faith.

The ideal scientist

This chapter has endeavoured to describe some of the characteristics of biology as a science. Biology deals with ideas, and these ideas are of three main kinds. **a** Some are general beliefs about the way nature is organised; **b** facts are collected by objective observation; and **c** theories are invented to explain these facts. All these interact with each other, but in particular the beliefs and facts are used to examine the theories. As a result, some theories are accepted and others rejected. This is called 'evaluating' the theories, and it is the subject of the next chapter.

Biologists vary as much as any other group of people. Even the greatest rarely excel at all the skills we would expect to see in the ideal scientist – Superscientist. If you were Superscientist you would be an awesome figure. You would have an intuitive ability at choosing problems, and a fluent imagination for thinking up theories. You would be skilled at mathematics and logical argument, in order to work out the details of a theory. You would also be an accurate observer and ingenious experimenter. You would be persevering when up against difficulties, but would know when to abandon a theory if it failed testing. You would be absolutely honest, and not least you would be able to speak and write clearly in order to communicate your findings. The kind of scientist you become will be largely a matter of which skills you are best at.

2 Judging theories

Values

Evaluation means deciding the value of something. We might say a new electron microscope is valuable and an old light microscope with cracked lenses is worthless. Although these are evaluations, they are not very useful. The reason is that one person may be thinking of the amount of money that could be gained by selling them, while another could instead be deciding how useful the instruments are.

There are several kinds of value, and five are listed in figure 4. You may be able to think of others. What they all have in common is that we prefer highly valued things to others with a lower value. But as we have seen, different kinds of value are easily confused because sometimes the same word is used for more than one. Obviously, to think clearly about evaluation, we ought to know what kind of value is being judged. You may like to try your hand at it: suppose a teacher says 'Good!' when a student gives an answer to a question. Is this evaluation? If so what kind of value is it, and who or what is being judged?

This chapter is concerned with evaluating biological theories. It means looking into the methods biologists use in deciding whether to accept or reject a theory. That such evaluation is commonly performed is clear enough; of the multitude of theories that have been proposed during the history of biology, only a few are accepted at the present day. How many people, for instance, now know anything about Darwin's pangenesis theory, let alone accept it?

Some people speak about 'the' scientific method, suggesting that there is only one method of evaluation. In fact there are several, and in the rest of

Figure 4 Five kinds of value, with adjectives used to describe them. Three of these values are used to judge biological theories.

Value	High value	Low value
Monetary	Valuable, expensive	Worthless, cheap
Aesthetic value or beauty	Beautiful	Ugly
Utility or usefulness	Useful, valuable	Useless, NBG
Moral value	Good, right	Bad, wrong
Truth	True, right	False, wrong

the chapter some of them will be described. These are mostly methods of judging the truth of a theory, but two other kinds of value are sometimes used, and will be mentioned at the end.

Testing predictions

A well-known method is the **hypothetico-deductive** approach. Although rather a mouthful, this term is useful because it mentions two of the steps in the procedure. First, one must have a theory or hypothesis. This must be sufficiently precise for **predictions** to be made from it. These predictions are potential observations that are derived from the hypothesis by deduction. Thus the hypothesis comes first, then deduction yields a prediction, and finally the prediction is compared with actual observations.

Consider an example from biogeography. This is the study of the way animals and plants are distributed around the world. Particular attention has been focused on the number of species found on different islands. According to one hypothesis, the number of species (S) found on any island in an archipelago is related to the area of the island (A) by this equation:

$$S = cA^z.$$

c and z are constants. From this it can be deduced that, when the logarithm of S is plotted against the logarithm of A, the scatter of points should form a straight line. This hypothesis, incidentally, is only a generalisation; it does not attempt to explain anything.

Figure 5 shows three possible results. Graph **a** would be taken as evidence for the hypothesis. Admittedly the points do not all lie exactly on the straight line; a certain scattering is to be expected on account of

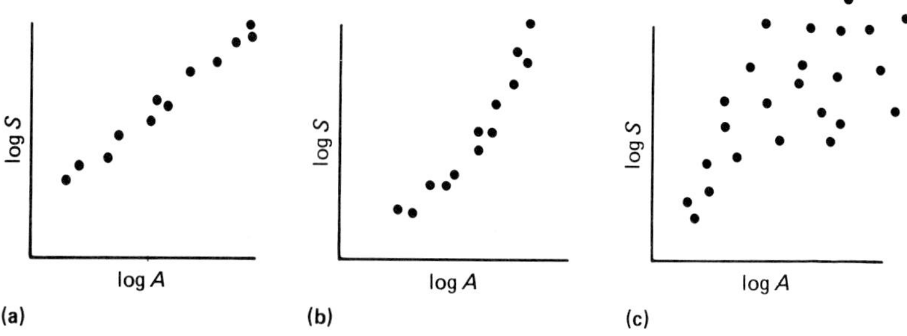

Figure 5 Testing the hypothesis $S = cA^z$: **(a)** is evidence for the hypothesis, while **(b)** and **(c)** are evidence against.

random variation (see p. 33). Graphs **b** and **c** are nothing like straight lines, and so are evidence against.

These two kinds of evidence, for and against, are not equally conclusive. The evidence against is much more powerful than the evidence for. This follows from the fact that the equation is supposed to be true for *any* group of species in *any* archipelago. As soon as one example is found which does not fit the hypothesis, the whole hypothesis is disproved. In this particular case, many such examples are known, so the hypothesis is untrue.

But why is the favourable evidence in graph **a** not just as powerful? Well, imagine that hundreds of studies have been made and each time a clear straight line has been produced. For instance, evidence for the hypothesis might have been found in the plants of Hawaii, the mosses on the Faroes, and snails in the Orkneys. Nevertheless, it is always possible that the next study will produce evidence against the theory. Perhaps the Darwin finches in the Galapagos will give a graph like figure 5**c**. So, however much favourable evidence is found, the hypothesis can never be conclusively proved.

When a hypothesis has a lot of evidence for it, and just one or two sets of evidence against, the latter are called anomalies. One significance of anomalies is mentioned in chapter 1 (p. 7): they can give the biologist a problem to solve.

The important thing about testing predictions is that they must be deducible from the hypothesis in a rigorous manner. On the other hand, complete deductions are hardly ever printed in the biological literature; indeed, to be presented fully they would usually take up a good deal of space. Admittedly, quite long deductions can be expressed in the form of algebra; but in non-mathematical areas of biology, hypotheses are often stated so unclearly that prediction is difficult or impossible. Then biologists often resort to the next method.

Plaice migration

Studies on the migration of plaice provide an illustration of this method. It is called 'testing expectations', and is described in the next section.

Staff at the Fisheries Laboratory, Lowestoft, have developed various acoustic transponding tags which can be attached to fish before they are released in the sea. An individual fish can then be followed by the research ship for many hours, during which time much is learned about the animal's behaviour and movement. The ship carries a sector-scanning sonar for locating the fish, and when a sound impulse reaches the fish from the sonar, the acoustic transponding tag on the fish transmits a pulse at 300 kilohertz (kHz) which is received by the sonar. In this way the fish

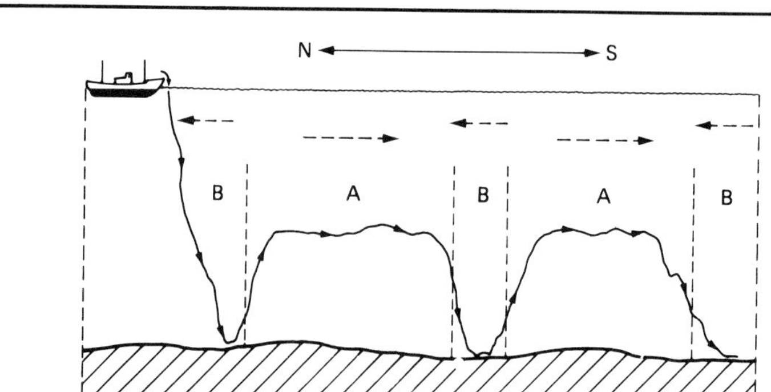

(b)

	Autumn	Winter
Number of paired tows	39	21
Tow pairs in which more plaice were caught on northward tide	5	20
Tow pairs in which more plaice were caught on southward tide	33	1
Tow pairs with equal catches	1	0
Total plaice caught on northward tides	64	180
Total plaice caught on southward tides	245	32

(c)

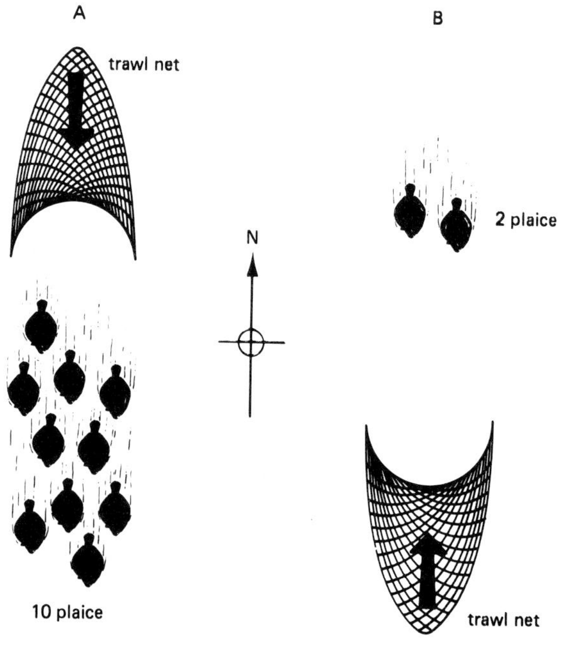

A

B

trawl net

2 plaice

N

10 plaice

trawl net

shows up clearly on the sonar display screen. The sonar enables the depth and plan position of the fish to be fixed. A more elaborate tag carries a compass, so that the direction in which the fish is heading can also be transmitted to the ship.

After tracking on separate occasions at least twelve plaice, a fairly clear pattern emerged. Over the course of a day or more, most of the fish covered a good distance in a fairly steady direction. The way they did this was not to keep swimming doggedly on in one direction. Instead they made efficient use of the tidal streams which flow first in one direction then the opposite way in alternating six-hour (h) periods. A fish stayed on the bottom during the tide which flowed against the fish's direction of travel. Then during the opposite tide it came off the bottom into midwater and travelled quite fast, mostly drifting with the tidal stream. So, a southward-moving plaice made use of the southward tides but hardly changed position during the northward tide (figure 6a).

The main interest of this discovery was that it prompted a theory about how plaice undertake their lengthy migrations. These occur, twice a year, in a north–south direction past East Anglia between the feeding and spawning grounds. Clearly, the behaviour of the tagged fish did not prove how plaice generally migrate. For instance, their behaviour might have been affected by carrying the tag, by the sonar waves, or by their prior captivity. But it was reasonable to suggest the theory that plaice on their autumn southward migration to the spawning grounds are in midwater during the southward tides and on the bottom betweenwhiles; and that the northward-migrating fish in the winter are in midwater during northward tides and otherwise on the bottom. The question then was: how could this be tested, preferably using fish that had not been caught and tagged?

Roy Harden Jones and his colleagues at Lowestoft decided that a good test would simply be to go fishing, using a trawl towed along in midwater. The trawling periods were arranged in pairs, one while the tidal stream flowed in one direction and the other while it flowed the opposite way. Thus it was hoped to compare the number of plaice caught in the two

---

Figure 6 Plaice migration. **(a)** Travel relative to the sea bottom of a plaice released from the research ship. The horizontal dimension of the diagram represents distance in a north–south direction. Each section A is a six-hour period when the tidal stream flows southward, indicated by the interrupted arrows. Similarly, the stream flows northward in each section B. **(b)** The results of testing the plaice-migration hypothesis (Jones *et al.*, (1979) *Journal du Conseil international pour l'Exploration de la Mer*, **38**, 331–337.) **(c)** This shows how the tidal-stream transport theory is compatible with fewer plaice being caught in autumn on the southward tide (A) than the northward tide (B). The net was towed through the water at 3–4 knot or 150–200 cm s^{-1}. Plaice can swim at a speed of at least 2.5 knot.

tidal streams. From the theory about the migration, it was expected that in the autumn more plaice would be caught in the southward tidal stream than the northward, while the reverse should happen in the winter. This indeed was found to be true (figure 6b).

Testing expectations

Satisfactory though the result is, what kind of test is it? In particular, has a prediction from the hypothesis been tested? The answer is 'no'. The method of testing predictions is so valuable because a falsifying observation contradicts the theory. If we are certain of the observations, the theory must be rejected. The best way of distinguishing between these two types of test is to ask what you would do if a falsifying observation were made. Suppose that more plaice were caught on the northward tides in autumn than the southward; would you abandon the theory of selective tidal-stream transport? If I were Harden Jones, I should say not.

I might suggest that perhaps the trawl was not taking a constant proportion of the fish (figure 6c). For instance, actively migrating plaice, travelling south in autumn on a southward tide, might be alert and fast enough to avoid the trawl, which was towed southward. This would mean that catches were poor. Indeed, if the trawl and the plaice were travelling through the water at the same speed, none would be caught at all! On the other hand, the ones caught on the northward tide might also have been heading south. The trawl was towed northward, so that good catches might have been obtained even if there were few fish swimming in midwater.

The speed of the net in relation to the fish is only one thing which could be queried if the observation appeared to contradict the theory, so it would not be wise to abandon the theory too soon. In fact, to predict the expected observation, one would have to assume a lot more than just the theory of selective tidal-stream transport. One would have to make other assumptions about the efficiency of the trawling, such as the fishes' response to the noise of the ship and to the approaching net. Because the expected observation cannot be predicted from the theory alone, it is better to call it merely an **expectation** and not a prediction.

Compared with testing predictions, the use of expectations gives a much less rigorous method. It is not as serious if the actual observations are unexpected, because it might be one of the assumptions that is incorrect, not the theory. At the same time, the evidence gained by making expected observations is not as strong as a confirmed prediction. For example, although the actual trawling catches (figure 6b) are encouraging, it is still possible that the theory of selective tidal-stream transport is largely incorrect if it is meant to describe how the majority of plaice migrate.

While the 245 plaice were being caught in autumn, simultaneous trawling on the bottom might have given catches of thousands. This might suggest that most migrated along the bottom and only a few used tidal-stream transport.

Altogether, testing expectations gives a much weaker kind of evidence than testing predictions, whether it is for or against the hypothesis.

Crucial experiments

So far it has been assumed that only one theory at a time is evaluated: but the drama of science sometimes consists in the conflict between two rival theories, and great excitement can build up when a **crucial experiment** is devised. The point of a crucial experiment is that it is designed to test two rival theories at the same time, so as to establish which is the winner and which must be discarded. On the other hand, the really unexpected might happen if the observations contradict both theories.

An example of a crucial experiment is one performed by Meselson and Stahl. The rival theories concerned how DNA replicates. According to the semiconservative theory, the double-helix molecule splits into two strands, each of which then acquires a new partner. So, each molecule is now half old material and half new. The conservative theory, on the other hand, assumes that the old molecules remain intact, and that new molecules are made only of new material.

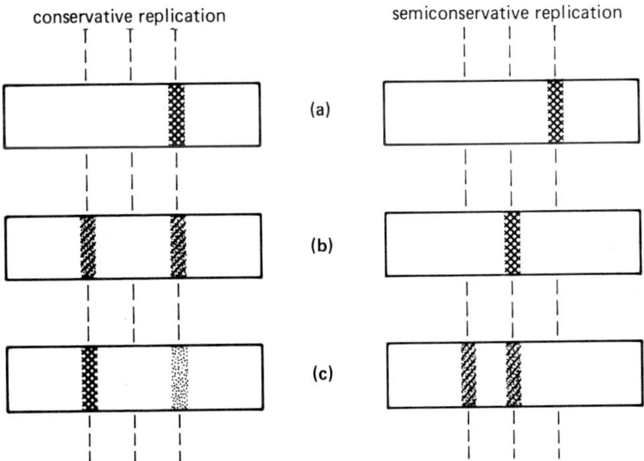

Figure 7 Meselson and Stahl's experiment: the patterns of DNA corresponding with the two theories of DNA replication. The heaviest DNA band in each tube is to the right, and the least dense to the left. **(a)** is the earliest and **(c)** the latest of the three observations.

Meselson and Stahl realised that if they introduced a particular heavy form of nitrogen into the new material, they could distinguish between the old and new material by their different densities. These densities would be measured in an analytic centrifuge. The patterns corresponding to the two theories of DNA replication are shown in figure 7. As a crucial experiment it worked splendidly, for the results unambiguously fitted the semiconservative theory and not the other.

Crucial experiments may be conclusive when the two theories give rival predictions. If only expectations are being tested the evidence is weaker, however clear-cut the observations. Was the Meselson–Stahl experiment testing predictions or expectations? To decide this we need to know what assumptions were made to work out the two sets of patterns in figure 7. The assumptions are built into the way the experiment was carried out and the way the centrifuge works. These questions are explored further in chapter 6.

Inconsistency

Accepting a contradiction is a logical crime. To do so would be to accept that a statement is true and not true at the same time. On the other hand, the rejection of contradictions is a powerful tool of thought. Many mathematical proofs are based on it, and we have already seen that it gives prediction testing its power.

The hunting down of contradictions has a wider application. It is not always necessary to make observations to evaluate a theory. Someone, without stirring from an armchair, might examine a theory and discover that it contradicts itself. Another way of saying this is that the theory is **inconsistent**. This would be a perfectly good reason for its immediate rejection. Any attempt to test it with observations would be a waste of time.

Inconsistent theories rarely get published, because any contradictions are usually detected very soon. Here, instead, is an imaginary example. Someone might suggest that the human species *Homo sapiens* is 20 000 000 years old, and that throughout that time its population has been increasing exponentially at a certain rate. Then another biologist might deduce from the theory that the population was zero only 15 000 000 years ago. This would obviously contradict the date given for the origin.

Finding a contradiction is a good method of evaluation. However, failure to find inconsistency is not in itself an adequate reason to accept a theory. Gobbledegook can be as free of contradictions as a true theory is.

Occam's Razor

Occam's Razor is a well-known and ancient rule for evaluating theories. It is named after William of Occam, or Ockham, or Oakham, who was an early fourteenth-century thinker living in Oxford. His rule states that where there is a choice a theory should be as simple as possible.

A variation of Occam's Razor is Morgan's Canon, introduced by the nineteenth-century animal psychologist Lloyd Morgan. Referring to animal behaviour, he stated that we should not interpret an action as the outcome of a higher psychical faculty if it can be interpreted as the outcome of one which stands lower in the psychological scale. It means that a theory representing an animal as a mechanical contraption is preferable to one recognising consciousness, purpose, or thought in the animal's mind.

This is rather hard on animals if they actually have those mental powers; and no scientist will ever be able to show that they do not. However, Occam's Razor and Morgan's Canon are sensible rules for biologists to follow. For one thing they make it easier to examine and test theories. Whether they are good rules for sorting out true theories from false ones is not so obvious.

In her book *An Imagined World*, June Goodfield describes the reactions of biologists to a certain theory about iron in the immunity system. Some rejected it because it was an unnecessary complication of another, simpler theory. Another biologist was suspicious of any simple explanation, if only because biological processes usually turn out to be very complicated. Indeed, Occam himself did not think that nature was necessarily simple, merely that we ought to use only simple theories.

Accordance with accepted theory

Chapter 1 makes a distinction between scientific theories and biologists' general beliefs about nature. A biologist formulating a theory will usually bring it into line with his general beliefs at an early stage. One of these beliefs is that observable facts are not the result of miracles or other supernatural causes. So, even if a biologist thinks a miracle is a possible cause of something, such as the origin of life on the earth, the idea is unlikely to be incorporated into any theories he thinks up.

Sometimes these general beliefs are so firmly held that it is impossible to imagine they are wrong. Can *you* really imagine a miracle occurring here and now in the twentieth century? I find it very difficult, not so much because miracles are impossible, but because my imagination is limited. This means that one's inadequate power of imagining things can evaluate theories by preventing them from being taken seriously.

The 'accepted theory' we have so far thought about has been scientists' general beliefs, but sometimes a theory may be evaluated by how well it fits with other scientific theories. An example of this occurred in a debate following Denys Tucker's controversial suggestion about the breeding of European freshwater eels (figure 8). The orthodox opinion was that the eggs hatch where they are laid in the Sargasso Sea, between Bermuda and the West Indies. Then the young eels take three years drifting in the ocean currents before reaching Britain and other parts of Western Europe. On arrival they ascend rivers to feed, eventually descending to the sea once more. Finally they swim or drift back to the Sargasso to spawn.

Tucker suggested on the contrary that when European eels descend to the sea, they are actually spent and simply expire. They never reach the Sargasso and do not reproduce. According to this idea, European eels *Anguilla anguilla* are really the same species as the American population *Anguilla rostrata*. They are simply a wasted part of the population which never contributes any genes to the next generation.

The theory provoked hostility because it painted a picture of an untidy, wasteful and pointless process. It was also said to conflict with the theory

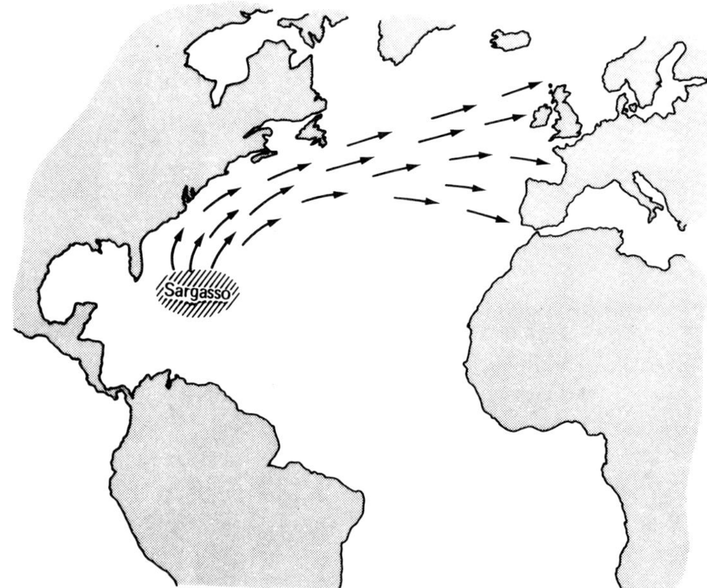

Figure 8 Route followed by eels, according to Tucker's theory: the young eels reach Europe from the Sargasso Sea, but do not return to breed.

of natural selection. Surely, it was argued, natural selection would gradually eliminate those eels (and their genes) which persisted in producing offspring that had no chance of reproducing. You might like to decide whether or not this argument is valid. People who accepted it naturally thought that it was powerful criticism of Tucker's theory.

There is a danger in accepting new ideas only if they fit in with already accepted theory. This is because the already accepted theory may not be true. Even the most apparently certain theory may be overthrown someday. For instance, there was nothing more certain in the nineteenth century than Newton's theory of gravitation, but that crumbled when Einstein challenged it. The danger in using this method of evaluation is that a whole complex of theories can be built up, depending for their acceptance on one original theory. If someone later discredits that original theory, the complex edifice is liable to collapse. Everyone who supported it then looks pretty silly.

Rejection of alternatives

The importance of having a good imagination plays a part in another method of evaluation. An unusual feature of this method is that the theory being evaluated is not itself tested. It consists in showing that several theories to explain a certain phenomenon are untrue; therefore, the only other possible theory must be true. If the theories considered are indeed the only possible ones, it is a sound and useful method. The trouble is that one cannot always be sure one has thought of all the possibilities. Take, for instance, a debate on the regulation of populations that took place in the 1960s. One set of biologists declared that, since herbivore populations are not kept within bounds by food or the weather, they must be limited by predators. Noticing this, a critic said that on the contrary there were other possible factors: populations of herbivores could perhaps be limited by shortage of space or nesting sites.

The method has commonly been used, sometimes incorrectly as we have just seen. In the nineteenth century, Adam Sedgwick gave his opinion on the superficial geological deposits known as 'drift' : they must have been caused by a great Flood, because there was no other possible cause. However, now it is generally believed that the cause was icesheets. Even today we still sometimes hear the argument that species must have originated through evolution because the only alternative theory, that God created them supernaturally, is untenable. Do you think this argument is conclusive?

Other values

This chapter began with a list of values, and so far we have mostly considered truth value. A possible exception is Occam's Razor, for it is not obvious that simple theories are more likely to be true than complicated ones. Perhaps Occam's Razor is not concerned with truth so much as with aesthetic value. It is well known how mathematicians can go into raptures about the elegance of a mathematical argument. For biologists, too, it can be a real pleasure to see a jumbled set of facts fall into place when related to each other by a simple theory.

Another value is utility, and the adjective often used is 'fruitful'. A theory is said to be fruitful if it suggests new problems or directs biologists' attention to new areas of research. So, a theory might be valuable in this way even though it is known to be untrue.

One reason for describing all these methods of evaluation is to help you decide for yourself about any particular theory. Some theories may be clearly acceptable or unacceptable on the basis of a powerful method of evaluation. Inconsistency and testing predictions are both strong methods. With another theory, one might not be able to use these methods. Then biologists will accept or reject it on much weaker grounds, such as by testing expectations. A possible example is the theory of evolution: it is difficult to see how it can be evaluated by any of the powerful methods. Today there is a widespread attitude among biologists that one should make up one's mind about important theories, even if there is no strong way of evaluating them. On the other hand, there may be something to be said for keeping an open mind.

Evidence and explanation

This book is all about observation, so it is worth noting which kinds of evaluation involve observation. Of the methods described, observation is essential in two. It can provide powerful evidence for or against a theory by testing predictions, and weaker but often important evidence through testing expectations. It might also play a part in the process of rejecting alternatives and in deciding whether a theory is in accordance with already accepted theory. It is not directly relevant to inconsistency and Occam's Razor. **Evidence** consists of any scientific observations used in any of the methods of evaluation.

An example of evidence was given earlier, that is the greater number of plaice caught on the southward tide in autumn. It is evidence for the selective tidal-stream transport theory. At the same time, the theory **explains** the observation. Often, evidence and explanation are just two sides of the same coin. However, an explanation does not always mean that the thing explained is useful evidence for the theory. For instance, we could say that the Darwinian theory of evolution explains why there are as many as thirteen species of Darwin finch on the Galapagos islands. Equally, someone might say that they all flew to the islands, and that this explains the facts too. Clearly, the facts are not much use as evidence. They do not even help decide between these two very different theories, one of which must be untrue. In fact, one can only judge how good evidence is by studying the method of evaluation used.

3 Observation

Science revolves around facts. If biologists disagree about the truth of a theory, they will usually try to find facts relevant to the dispute. Observed facts are particularly important, because they have a stability and dependability which is at the heart of the scientific enterprise. Theories may come and go, but facts go on for ever.

The special role of observed facts is sometimes put like this: facts are objective, while theories are matters of opinion and must be evaluated. But, this is too simple, for scientific facts always involve some assumption or interpretation, even if it is not obvious. Our eyes may receive information about the intensity and colour of light from numerous minute spots on a sheet of paper. These are the **sense data**. However, what we are conscious of is something much simpler. In figure 9a, the pattern could be described as an inverted white triangle covering two upright triangles. We could even measure the white triangle and work out its area, but we would be wrong if in fact the white shape is made up of three stars.

To take another example, someone might show us a photograph that appears to be the Queen's head on a coin (figure 9c). Most of us would say that that was what it was. But it is quite easy to fool the observer; one way is by using a photograph of the impression the coin makes when pressed into wax, and lit from the other side. The two images are virtually

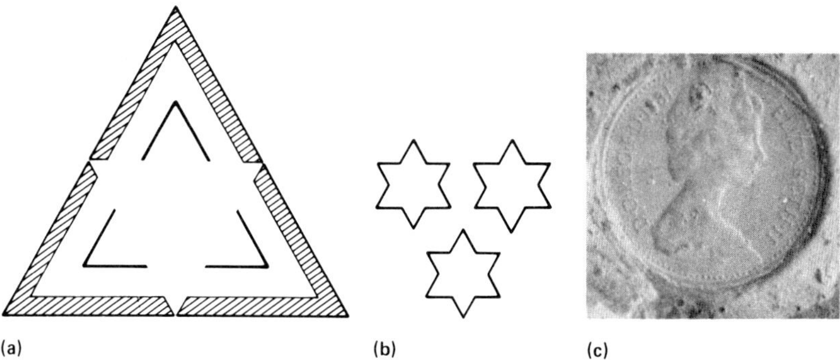

Figure 9 **(a)** This pattern of dark lines would normally be seen as a white triangle covering two others. **(b)** The white shape in **(a)** might instead be three stars. **(c)** Is this a coin, or the impression of a coin in plaster or wax? (Photographs can be printed backwards!)

identical though they are of quite different things.

This shows how we usually interpret sense data in terms of simple or familiar ideas. We communicate our observations to other people by using these interpretations. We do not give a table of the intensity and colour of millions of spots which together make up the image.

Optical tricks such as these may seem a little removed from practical science. Instead, suppose that, in a study of geographical variation in hares, we find that adult hares in a certain area have ears of average length 7.0 cm, standard deviation 0.6 cm. The observation may not fit in with someone's theory, and so he may object. Perhaps the hares were all measured at a time of year when they had not finished growing: or perhaps the sample was not large enough. A really determined objector could argue that whenever a hare is measured, a nervous reaction in the animal makes the ears shrink a little. If this were true, measuring the ears of live hares would be almost impossible.

Objectivity

Carefully observed facts are accepted in science with a certainty not granted to theories. Only a crank or a troublemaker would go to great lengths to dispute the length of those hares' ears. This gives scientific observations a very special status. Facts are the ultimate authority. This is sometimes expressed by calling them '**data**', which is Latin for 'given'.

What then does it mean to say facts are **objective**? It merely means that if any other competent scientist followed the same procedure, the same observations would be made. Anyone counting the wings on a normal dragonfly agrees that there are four. This is an objective observation. However, two people listening to a record may make different remarks about it: one thinks it is fantastic and the other thinks it is terrible. These are not objective observations. It is because scientists agree about objectively observed facts that they can be used to arbitrate when evaluating theories.

Important observations may be ignored for a time but are rarely rejected permanently. In 1944 Oswald T. Avery published the results of an experiment; this seemed to show that purified DNA from a donor bacterium, when added to a culture of another strain, converted some of the latter into the donor strain. The discovery was astonishing at a time when the hereditary material was thought to be protein. Nevertheless, important though the facts were, they were largely ignored until other experiments in 1952 pointed to the same conclusion. With hindsight we can say that the observed facts were not properly used to evaluate the theory that protein carries the genetic information.

Observation in research

It was common in the nineteenth century for scientists to say that science starts with observations. Nothing can happen until the facts are gathered, for only then can a theory be found to explain them. With almost the same unanimity, recent textbooks have placed observation later in 'the' scientific method. Hypotheses must be proposed, predictions deduced, and then observations made to test the hypothesis.

Both these views are oversimplifications. In this book it is assumed that there is no single scientific method. Observation can be performed at various stages in a piece of research.

It is useful to distinguish between casual and systematic observation. Some research projects start when a biologist notices something by chance. A casual observation of this kind was made some years ago on Bardsey, a small island off the Welsh coast. The commonest slug there is polymorphic: one form is entirely black while another is bright yellow or orange with a black stripe along the back. It was noticed that the ratio between the numbers of these forms varied from one part of the island to another. Little more could be done until further observations were made, so a plan was drawn up to measure the ratio at a large number of points on the island. This second stage of observing can be called systematic because some plan or system is followed.

Only then, with a mass of systematic observations, was it sensible to think up a hypothesis. At this stage a third set of observations was needed, to test the hypothesis. These were also systematic, because a plan was used. This time the plan meant working out predictions or expectations from the hypothesis and making observations to see if they were correct.

The first two kinds of observation are not essential in biology. It is explained in the first chapter how theories can be invented using no observations at all. However, the third kind is much more important because it helps decide whether a theory is true or false.

Operational definitions

Working out what observations will help evaluate a theory is not always easy. There may be no problem when the hypothesis is about blood temperature, a flower's colour, or the size of a flea population in a nest. Observing all these is straightforward. But what about hypotheses concerning the evolution of species, trophic levels in an ecosystem, or the aggressive behaviour of blackbirds? Although this list may seem like the first one, there is something special about evolution, species, trophic levels, ecosystems and aggressive behaviour.

For instance, to which trophic level does an insect belong which feeds both on detritus and living plants? Or if a blackbird makes a half-hearted lunge towards a dead snail, is that aggressive behaviour, an expression of curiosity, or indecision about whether to eat it?

A hypothesis dealing with these vague but important ideas is difficult to test unless we know precisely how they can be observed. Another way of saying this is that biological concepts need an **operational definition**. We cannot directly measure a bird's kinetic energy, but its mass m and velocity v can be measured, so that the kinetic energy can be calculated. Kinetic energy has an operational definition $\frac{1}{2}mv^2$ that no one disagrees with.

Something like a trophic level gives more problems. There is no obvious way of converting observations into statements about trophic levels, so an arbitrary decision may have to be made. For instance, an insect may be included in the trophic level which represents its major energy source. On the other hand, it might be better to have a more complicated operational definition. Suppose the insect spends two thirds of its time as a detritus feeder and the rest as a herbivore: then 0.67 units could be added to the one trophic level and 0.33 to the other.

Controlled experiments

The design of experiments deserves a whole book; here only a few points are made to emphasise that experiments are just one way of making observations.

Experiments involve setting up an artificial situation before making observations. The most extreme kind is the **controlled experiment**. Several treatments are set up, but all the conditions are kept the same except for one. Then the differences in the measured effect are due to the one condition that is varied.

A complicated form of controlled experiment is the **factorial experiment** (figure 10). Here two or more conditions are varied, and much more information is gained than from a simple controlled experiment. In fact, it consists of several controlled experiments performed at the same time.

Figure 10 The plan of a factorial experiment on the growth of plants. It consists of twelve separate treatments, numbered 1 to 12 in the table.

			Light intensity		
		A	B	C	D
Temperature	X	1	2	3	4
	Y	5	6	7	8
	Z	9	10	11	12

Thus treatments 1, 5, and 9 are one experiment; the temperature is varied and light intensity is kept constant. Each column and each row of the table is one controlled experiment.

It is not always possible to make an experiment controlled. For example, one may want to grow seedlings in a range of potassium solutions, keeping all other conditions constant. But a lower concentration of potassium ions also means that the osmotic pressure is lower: or if another ion is added to keep the osmotic pressure the same, then there is a higher concentration of that ion. Varying only a single condition is sometimes impossible. The controlled experiment is an ideal, and not always practicable.

Circumstantial observation

At the other extreme from controlled experiments are **circumstantial observations**. An example might be noticing the blackbird's behaviour when you are out bird-watching. Between these extremes there are intermediates. For instance, a natural fire in savannah usually produces a sharp boundary in the vegetation where the fire stops. A similar effect might be caused by an abrupt change in the geology of an area. Situations like this are sometimes called 'natural experiments'. A hill farmer may fence off a few hectares of moorland, with the result that different kinds of grazing on the two sides of the fence produce vegetation changes. This is an 'unintentional experiment'. Another could result from a fire started deliberately. Crossing two strains of fruit fly to observe the progeny is an uncontrolled experiment.

Comparing the extremes, each kind of observation has its good and bad points. Experiments allow conditions to be carefully regulated and precise measurements to be made. They also allow the organism to be subjected to a much wider range of conditions than occur naturally. However, the artificiality of the situation has disadvantages. Intensive preparation of tissue or cells may mean that the observations bear little resemblance to what goes on in living organisms. This is illustrated in the chapter on electron microscopy. Even if a living organism is being used in an experiment, the artificial surroundings may influence the way it responds. A monkey may have different food preferences in the top of a tree in Kenya than when faced with the same choice in glass dishes at a zoo.

Circumstantial observations can often give information much more relevant to the organism in the wild. One drawback is, of course, that there is much less control of the environmental conditions. Another is that observations may have to be made in difficult circumstances. Measuring something to three significant figures up a gum tree may not be as easy as in the laboratory.

All in all, where practicable, both experimental and circumstantial observations should be made. In this way the advantages of both methods can be gained. You may remember the toadstool example in the first chapter (figure 1). In this kind of study it is best to do experiments in the laboratory, such as measuring the distance spores are carried in wind of different speeds, and also to make measurements in the field. These could be of the wind speed at different heights in the grassland, the time of day the spores are released, and so on.

Qualitative observation and classification

Quantitative observation involves measurement. Important though this is, biology can achieve a great deal without it. However, without qualitative observation biology would have got nowhere. This needs to be explained.

Suppose a large number of fish are caught from a lake and weighed. A good way to summarise the results is to plot a frequency-distribution diagram. Two possible results are shown in figure 11. In **a** the mass shows continuous variation, but in **b** it is discontinuous.

In either case we can refer to each fish by its mass: such and such an individual is 0.62 kilogram (kg). This is making a quantitative observation. But it might be convenient in the right-hand diagram to divide the fish into sets or classes, A and B. If we then say a certain fish belongs to class B, we are making a **qualitative observation** about it. It either does or does not belong to class B, and no measurement needs to be given.

What has been done is to make a classification. This is true even if groups A and B are merely the immature and adult members of one species. **Classification** means grouping similar things and separating dissimilar ones. Just about anything can be classified, quite apart from the

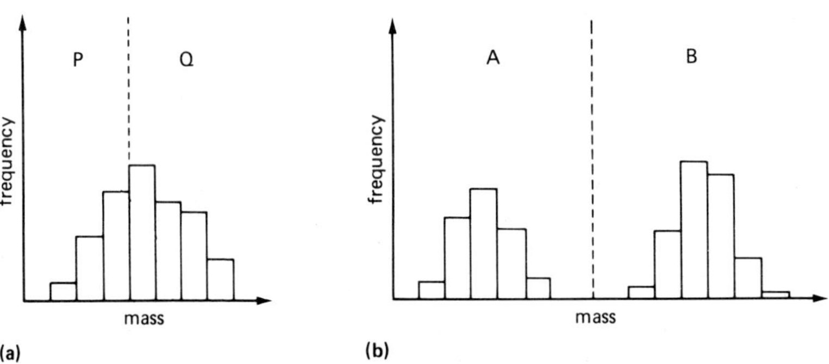

Figure 11 Frequency distribution histograms of the masses of fish caught in a lake. In **(a)** the mass shows continuous variation, but in **(b)** it is discontinuous.

species and genera of organisms that immediately come to mind. In biology we use classifications of blood cells, tissues, parts of the flower, chemical compounds, colours, behaviour patterns, species, instruments, and so on.

Classifications can be good or bad, and it is worth considering what makes a good one. It is possible to divide the fish in figure 11a into classes P and Q merely by choosing an arbitrary boundary, but this would not be a good classification for general purposes. For instance, it might be difficult to decide, in the case of fish close to the boundary, to which class they belong. One fish might land up in class Q merely because it was weighed just after eating food. On the other hand, the classification in figure 11b is a good one because there is no doubt where every fish belongs. The reason is, of course, that the boundary has been placed in the gap, at a natural discontinuity.

So, a natural discontinuity is needed for a good classification. Often though, there are no clear discontinuities, at least not in a single character such as the fishes' mass. But when several characters are taken into account, useful discontinuities may be discovered. In figure 12, for example, four spiders show each degree of variation in both body shape and pattern. This is evident by counting the spiders in each column and each row. But there is also a clear discontinuity, and so it makes sense to group the spiders into two classes.

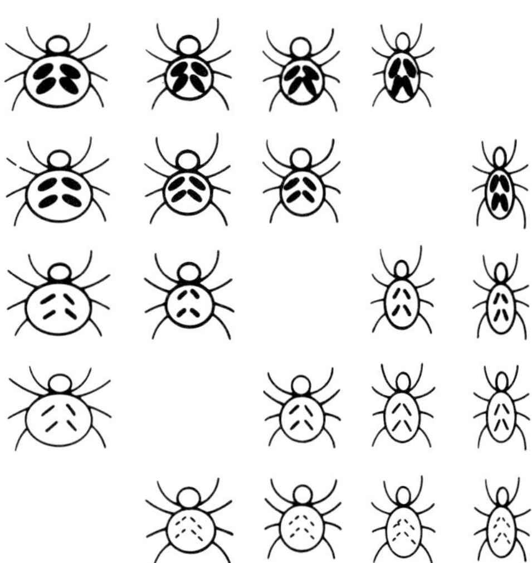

Figure 12 Variation in spiders. The character varying up and down the diagram is prominence of the pattern. Left to right, the variation is in body shape.

Some classifications of species are much more complicated than the classification of these spiders. Designing classifications is often difficult and the taxonomists who classify species are highly skilled biologists. The sheer difficulty of some classifications means that other biologists need help in identifying organisms. Usually this help is provided in the form of identification keys. These are discussed in another book in this series, *Mathematics in Biology*, chapter 6.

The role of classifications

Why is classification so important? The reason is that we need it to make any observation, even measurements. This should be clear from two examples. Remember the bird-watcher who wrote in his notebook 'The blackbird is behaving aggressively'. Three separate classifications have been used here. The bird has first of all been identified as a blackbird, rather than a thrush or a ring ouzel. The action has been identified as behaviour, rather than a passive response such as being knocked over or falling down dead. Finally, it is identified as aggression rather than curiosity or indecision. In each case an all or nothing decision has been made.

For a second example, suppose we record that a fish has a mass 0.62 kg. Two classifications are involved. The organism is identified as a fish, and mass in kilograms is chosen from a classification of variables in order to measure the fish. No measurement can be recorded unless the thing measured is identified and a variable is chosen. If you write nothing but '0.62' on a postcard and send it to another biologist, it will not mean much to him.

No scientific observation can be expressed simply in numbers. Some kind of language is needed. The importance of classifications is that they supply all the important words used in scientific observation.

Measurement

Two aspects of measurement need to be understood: precision and accuracy. When repeated measurements of the same thing are made, they will rarely be identical every time. The **precision** refers to how tightly clustered they are. A thermometer may give a set of measurements of a pond's temperature as in figure 13a, curge B. In contrast, a thermistor might give a more tightly grouped set of measurements, as in curve A. The smaller the standard deviation, the more precise the instrument is.
or sensitive the instrument is.

According to the thermistor the temperature might be recorded as 13.0 °C to the nearest 0.1 degree, while according to the thermometer

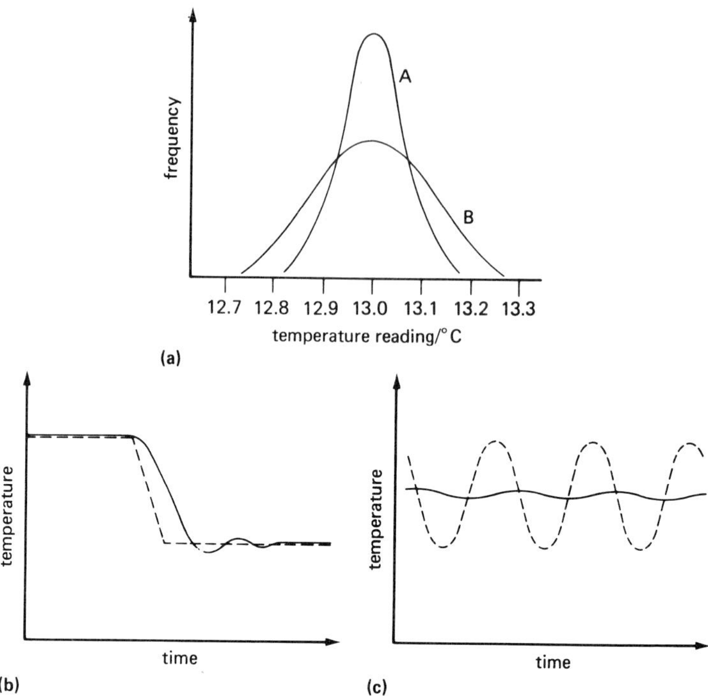

Figure 13 Measuring temperature. **(a)** Frequency distribution of the temperature measurements made on the same substance with (A) a more sensitive and (B) a less sensitive instrument. **(b, c)** In each case the real temperature is shown by an interrupted line, and the instrument's reading by a continuous line.

it is 13.0 °C to the nearest 0.2 degree. If the actual temperature then changed by 0.1 degree, the more precise instrument would probably detect it, but the less precise one which not.

So far the true temperature of the pond has not been considered. If it is exactly 13.00°C, then both instruments are **accurate**: but if it is 13.08°C, the less precise instrument is still accurate, because the real temperature is indeed 13.0°C to the nearest 0.2 degree. However, the thermistor would not be accurate, for the real temperature is not 13.0°C to the nearest 0.1 degree. It is 13.1°C to the nearest 0.1 degree. Sensitivity is expressed by the number of significant figures given in the measurement, or by the stated error. Accuracy is simply a matter of whether the measurement is correct.

At least, this is one way the word 'accuracy' is used, though you may find statisticians using it to mean something slightly different.

Errors

Even a completely accurate instrument would show the kind of variation illustrated in figure 13a. This is known as **random error**. Instrument designers try to make the random error small; but there is no point in reducing it so much that the instrument then becomes inaccurate.

Another kind of error appears when readings are consistently wrong in a particular direction. If a thermometer always reads too high, it may be because the scale is engraved in the wrong place, or because there is some other defect in the design or manufacture. These **systematic errors** make the instrument inaccurate. Biologists ought to check instruments for systematic errors before using them.

This kind of error can be easy to detect, but it is more difficult when the measured variable changes rapidly. For instance, the real temperature may suddenly decrease. The instrument might react quite quickly, but overshoot and oscillate before giving a new steady reading (figure 13b). Clearly, the instrument shows that the temperature changes, but it gives an inaccurate picture of how fast or steadily the change takes place. If instead the real temperature fluctuates rapidly (figure 13c), the instrument may not respond quickly enough. As a result it may give a more or less steady reading. Again, it is inaccurate.

There is yet another difficulty in any observation: the observer or the instrument may actually change the phenomenon being measured. In the eighteenth century Stephen Hales tied a horse to a gate, then fixed a vertical twelve-foot glass tube to the carotid artery by means of a goose's windpipe. The aim was to measure the blood pressure, and indeed the blood rose to nine and a half feet in the tube. But of course this treatment may have had an adverse effect on the physiological state of the horse. As a method of measuring the horse's ordinary blood pressure it is not ideal.

Any scientific observer needs to think very carefully about this problem. Is it possible, for instance, to measure how wide open the stomata are on a living leaf? There are several methods available, such as forcing air through the leaf, making a wax impression, using penetrating fluids, or looking at the stomata through a microscope. But can we be sure that there is a method which does not change the temperature, humidity, or other conditions around or in the leaf? It may in fact be impossible to measure how open a stoma is without changing it. One way round this difficulty is to make a copy of the phenomenon, for example on film, and then measure the copy. This is obviously useful in studying bird flight, since measurements are easier to make on a film and the bird is not disturbed. But photography may not solve the problem with stomata.

Instruments

Instruments are devices for making observations. They range from something as simple as a ruler to a complex machine like the electron microscope.

Nearly all biological instruments that come to mind are used for measurement. However, a few are used for making qualitative observations. Chromatography is a method which can be used to identify chemical compounds; it is described in chapter 5. Identification of species of organisms can also be difficult, and here again there are instruments to help. They are, of course, the identification keys. Although consisting of printed matter on paper or computer programs, they are used in the same way as ordinary instruments.

Often biologists using instruments do not understand in detail how they work. Instructions are followed and the results written down, but what goes on in between is a mystery. Instruments used like this are called 'black boxes'. But to use an instrument intelligently, one ought to have a reasonable understanding of how it works.

Many observations need no instruments. Even a babe in arms can distinguish tastes, temperatures, or the textures of things touched. In all these there is actual contact between the thing observed and the observer's body. But other observations involve transferring information from one point to another some distance away. Merely to see the shape of a plant, or to hear a bird singing, requires electromagnetic radiation or sound waves respectively. These transfer the information directly from the organism to our sense organs. When the information passes through an instrument on the way, this transfer of information may be much more complicated. To understand how an instrument works is largely a matter of understanding how information moves through it. The following chapters outline how this occurs.

This chapter has emphasised the central role of observation in biology. It is as important for the student in school as for the professional researcher. But making scientific observations is by no means easy. It requires an understanding of several ideas: the interpretation involved in any observation; the authority of facts in evaluating theories; the meaning of 'objectivity'; the role of observation at various stages of an investigation; the need for operational definitions; the respective good and bad points of experimental and circumstantial observation; the importance of qualitative as well as quantitative observation; an appreciation of the errors involved in observation; and an understanding of how instruments work. When you have mastered all this, you are well on the way to becoming a trained scientist.

4 Isotopic tracers

Tracers

When I was at school, photosynthesis was summed up in the following equation:

$$6CO_2 + 6H_2O \rightarrow C_6H_{12}O_6 + 6O_2$$

Little else seemed to be known about the reactions involved. How different things are today!

One of the early problems that needed a solution was this: did the oxygen given off come from the water or the carbon dioxide? If all oxygen atoms were identical in their chemical and physical properties, this would have been an extremely difficult problem. Fortunately, however, there are different kinds of oxygen atom, called **isotopes**. Although they are chemically identical, it is possible to distinguish between them physically. This means that the carbon dioxide can be made with one sort of oxygen, and the water with another. If the oxygen gas given off during photosynthesis is then analysed, it should be clear where it comes from. This was the experiment conducted by Ruben and Kamen before the Second World War. The two isotopes used were ^{16}O and ^{18}O. It was discovered that $^{18}O_2$ was produced only from $H_2{}^{18}O$; none resulted when $C^{18}O_2$ was used.

Many different atoms have isotopes that can be used in biology, so the powerful method established by this famous early experiment has been used now to solve many other problems in biochemistry. The principle is always the same. If we are interested in what happens to a particular atom in a compound, the compound must first be made with an unusual isotope in place of the atom. This is called **labelling** the compound. It can be done artificially or by getting an organism to do it for us. Then the compound is fed in to the reaction being studied. After that the reaction can be stopped at any time.

At this point the search for the label starts. It may not be attached to the original compound any more, so various compounds must be isolated and examined to see whether they have the label. Some of the techniques used to separate out the compounds are described in chapter 5. The present chapter is concerned with detecting the label and measuring how much there is. It will often be of interest to know more exactly where the atom has landed up; for instance, a carbon atom may be built into a glucose molecule, but which of the six carbon atoms is it? Finding this would involve purely chemical techniques which will not be described here.

Experiments using isotope labels to follow atoms from one compound to another are called **tracer** experiments. It is largely because of tracer experiments that we now have such an amazingly detailed knowledge of photosynthesis and many other processes.

Stable isotopes

The two kinds of oxygen, ^{16}O and ^{18}O, are called **stable isotopes**. This is simply because they do not change with time. A bottle of ^{18}O can be left on a shelf indefinitely without deteriorating. There are five isotopes of oxygen, but only three are stable. ^{16}O forms the bulk of the oxygen in the atmosphere and other materials.

Isotopes are known by their relative atomic masses. The ^{16}O atom is approximately sixteen times and ^{18}O eighteen times more massive than the hydrogen atom. The two isotopes are chemically identical. Since chemical properties are controlled by the electrons, this means that ^{16}O and ^{18}O have the same electron arrangement. In fact, all neutral oxygen atoms, that is oxygen atoms with no overall electric charge, have eight electrons.

So, the difference between the isotopes must lie in the nucleus. In a hydrogen nucleus there is only one particle. It is a proton, and it carries a single positive electric charge. This balances the negative charge of the electron to make a neutral atom. Similarly, each oxygen atom must have eight protons in the nucleus to balance the eight electrons. These protons

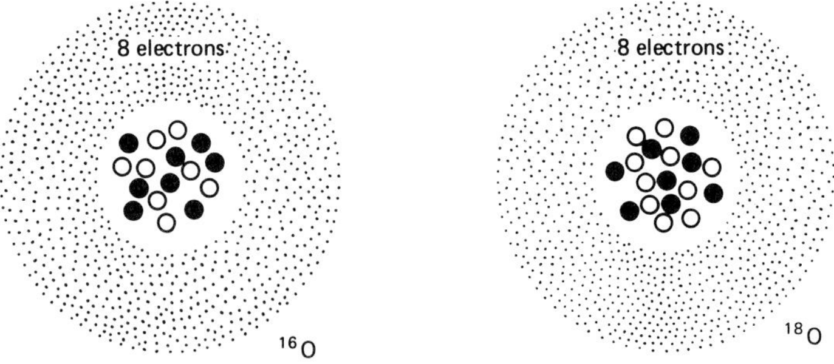

Figure 14 The two common oxygen isotopes. Each atom has eight electrons in the zone shown dotted, and eight protons (black discs) in the nucleus. The open circles are neutrons: eight in ^{16}O, and ten in ^{18}O.

account for eight of the units of relative atomic mass. This leaves eight other units in ^{16}O and ten in ^{18}O not yet accounted for. They are the particles called neutrons. Each has about the same mass as a proton, but no electric charge. It is the number of neutrons in an oxygen atom which determines which isotope it is (figure 14).

Time-of-flight mass spectrometer

The only property of stable isotopes that can be used to distinguish between them is their mass. This is no easy task. One method using the analytic centrifuge is described in chapter 6 (see p. 68). Otherwise the instruments employed are mass spectrometers.

Newton's theory of forces states that the force F on an object equals its mass m multiplied by the acceleration a caused by the force. In other words,

$$F = ma.$$

This gives a way of separating out a mixture of isotopes. If they are all subjected to the same force, the less massive atoms will accelerate away faster than the more massive ones. After a short while they should be moving in two quite separate groups. It is like a cross-country run, in which a group of good runners sets off at the same time as another group of mediocre ones. At the start they may be all mixed up in a large crowd, but gradually the good runners forge ahead until as a group they are well separated from the mediocrities.

The **time-of-flight mass spectrometer** works on this principle. Gas molecules are needed for the spectrometer, but they cannot be accelerated until they are given an electric charge. This means that they must be converted into ions. If one electron is knocked off each molecule, the molecule is left with a single positive charge. It is now an ion. The usual way to ionise a gas is to bombard it with a stream of electrons. An electron gun is used for this, like the one in an electron microscope (see p. 88).

At one end of the spectrometer is an ionisation chamber (figure 15a). A small volume of the oxygen gas is let in, and then ionised. The electrode in the chamber is maintained at a positive charge. Because unlike charges attract each other, this collects the spare negatively charged electrons knocked off the neutral gas molecules. At the other end of the long tube is another electrode, this one maintained at a strong negative charge. In other words it has a high concentration of electrons on it.

As soon as an oxygen molecule becomes a positive ion, it is attracted to the negative electrode at the far end of the tube. Because all the ions have the same charge, they are all attracted by the same force. They immediately shoot out of the ionisation chamber. The lighter ions are accelerated faster and reach the negative electrode before the heavier ones. If the tube is long enough, the two or three kinds of ion can be completely separated in this way. It only remains then to measure the relative numbers of each arriving at the electrode.

Each positively charged ion arriving at the negative electrode collects an electron from it and becomes a neutral molecule again. It is quickly removed from the spectrometer by the vacuum pump. If many millions of ions all take electrons from the electrode in a fraction of a second, other electrons move along the wire to the electrode to replace them. This of

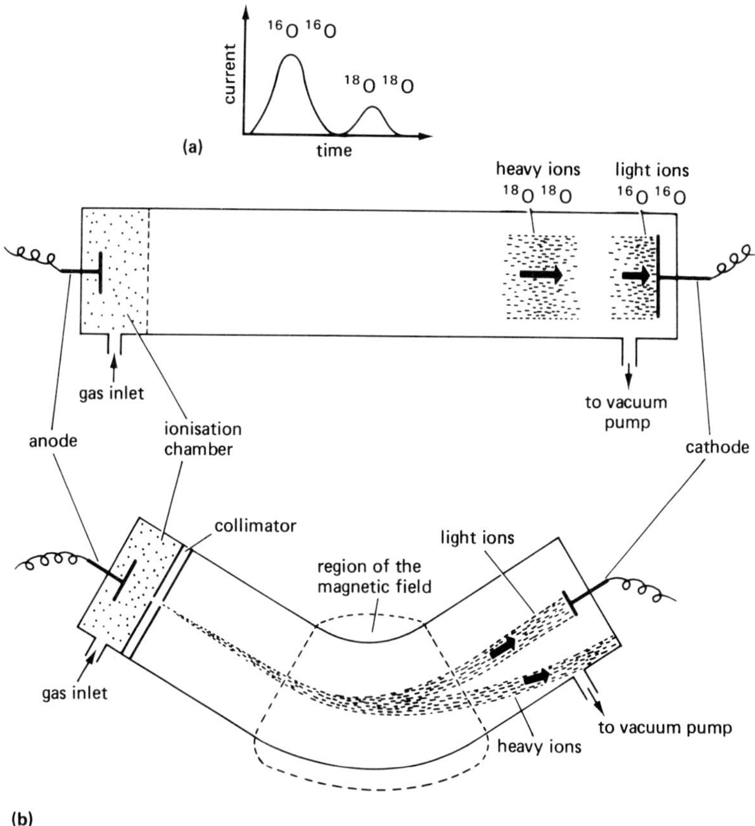

Figure 15 (a) Time-of-flight mass spectrometer. Lighter ions travel faster and reach the cathode before the more massive ions. The inset shows a record of the current as the ions arrive. **(b)** Magnetic-field mass spectrometer. The magnetic field changes the direction of the ion beams, particularly the lighter ions.

course is an electric current. So the number of ions hitting the electrode can be assessed by measuring the electric current flowing to it.

The graph in figure 15a shows the kind of record made by the $^{16}O^{16}O$ and $^{18}O^{18}O$ ions. As the $^{16}O^{16}O$ ions arrive, the current builds up to a peak and dies away. The same happens as the $^{18}O^{18}O$ ions take electrons from the electrode. Measuring the areas under the two curves gives the relative numbers of the two ions.

Magnetic-field mass spectrometer

Acceleration does not just mean changing speed in a straight line. It can also mean changing direction even if the speed remains the same. This is the idea behind the **magnetic-field spectrometer**.

A moving ion creates a magnetic field around it, because it is carrying a charge, so it will respond to another magnetic field it is passing through. The ion will experience a force, rather like a compass needle is forced to take up a certain position by the earth's magnetic field. In this spectrometer, a magnetic field is created halfway down the tube by a powerful magnet, and the ions have to pass through it.

The gas is ionised and the ions then shoot through a series of holes, the **collimator**. This makes them into a fine straight beam (figure 15b). They are attracted down the tube towards the negative electrode. When they pass through the magnetic field they are forced sideways. This makes them change direction. They all experience the same force, so the lighter ions change direction more than the heavier ones. The result is that the two kinds of ion are separated, forming beams one beside the other.

Because the two isotopes are not arriving at the end of the tube one after the other, gas can be fed into the ionisation chamber continuously. Then, by gradually changing the strength of the magnetic field, the beams can be swept slowly across the negative electrode. Measuring the current gives the same kind of graph as in the time-of-flight instrument. So, either instrument can measure the relative amounts of $^{16}O^{16}O$, $^{16}O^{18}O$ and $^{18}O^{18}O$ produced in the photosynthesis experiment.

Unstable isotopes

It was mentioned earlier that two of the five oxygen isotopes are unstable. These are ^{15}O and ^{19}O. If a bottle of ^{15}O or ^{19}O were left on a shelf, it would not keep for long. The isotope would decay and change into something else. This property of **unstable isotopes** might seem to be a nuisance, but in fact it makes them very valuable to the biologist. The reason is that they are much easier to detect than stable isotopes.

Unfortunately, oxygen and nitrogen do not have unstable isotopes suitable for biological experiments. However, most other biologically important elements do. A particularly useful isotope is ^{14}C. When a ^{14}C atom decays, it gives off an electron and turns into an ordinary atom of nitrogen, ^{14}N. The electron comes from inside the nucleus, in fact from a neutron which turns into a proton and the electron. It is because the number of protons in the nucleus changes that the atom turns into a different element (figure 16).

The electrons given off are known as **beta rays**. These can be dangerous, but with proper precautions ^{14}C is safe to use. Some other isotopes used in biology not only produce beta rays but **gamma rays** as well. Gamma rays are a kind of electromagnetic radiation (p. 115). They are much more

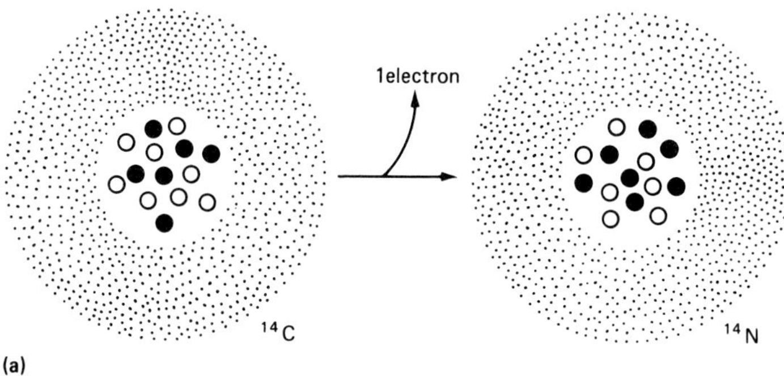

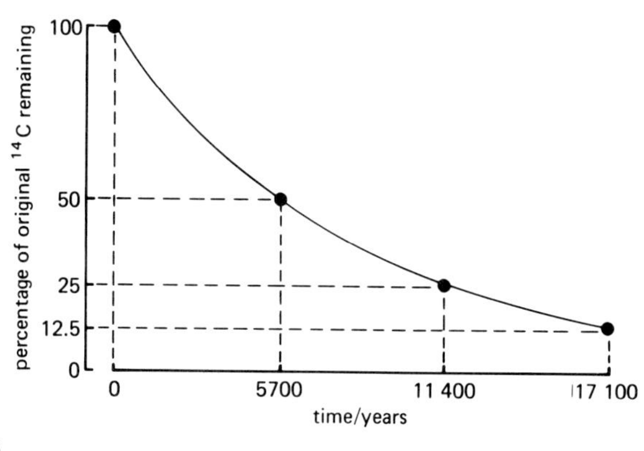

Figure 16 Radiocarbon. **(a)** A ^{14}C atom contains six protons (black discs) and eight neutrons (white discs). When it decays, one neutron changes into a proton. The new atom must be ^{14}N, since any atom containing seven protons is an isotope of nitrogen. **(b)** The decay of ^{14}C in an object. Comparison of the amount remaining and the amount assumed to be present originally gives the age of the object.

dangerous than beta rays, and so where possible biologists avoid using isotopes which produce them. Beta and gamma rays are two kinds of **radioactivity**.

Autoradiography

It is because unstable isotopes give out radiation that they are much easier to locate than stable isotopes. Using them, tracer experiments allow us to find the whereabouts of an atom in which we are interested. This often means following an atom through an organism or cell to see where it eventually gets used.

Autoradiography provides a neat way of finding the isotope. Just as light waves can affect a photographic film, so also can beta and gamma rays. So, the trick is to place a film on the material containing the isotope, and let the isotope take its own 'photograph'. In the example in figure 17,

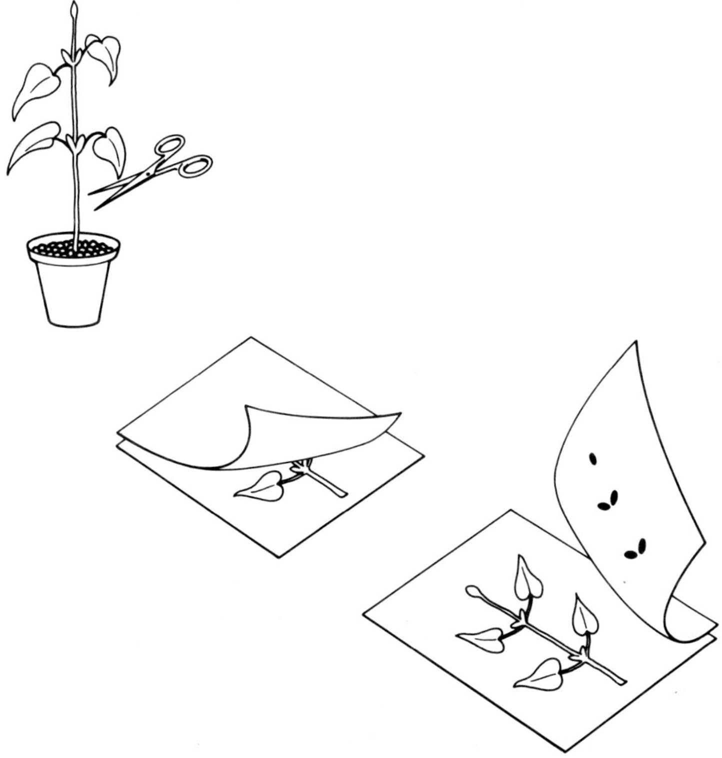

Figure 17 Preparing an autoradiograph. A plant or other material is laid on paper with film over it. When taken off and developed, the film bears an image of the parts the isotope has reached.

only the buds show up on the radiograph because that is where the isotope is concentrated. It is rather like making the buds light up in the dark, except that beta rays and not light rays are emitted.

The method is not restricted to whole plants or animals: it can be adapted to single cells. For instance, it provided useful evidence in the story of mitochondrial DNA.

DNA is often detected by using Feulgen stain, and claims were made in the 1950s that mitochondria took up Feulgen stain. The claims were not taken very seriously. Mitochondria were also isolated and DNA found in them, but it was thought to be contamination by DNA from nuclei.

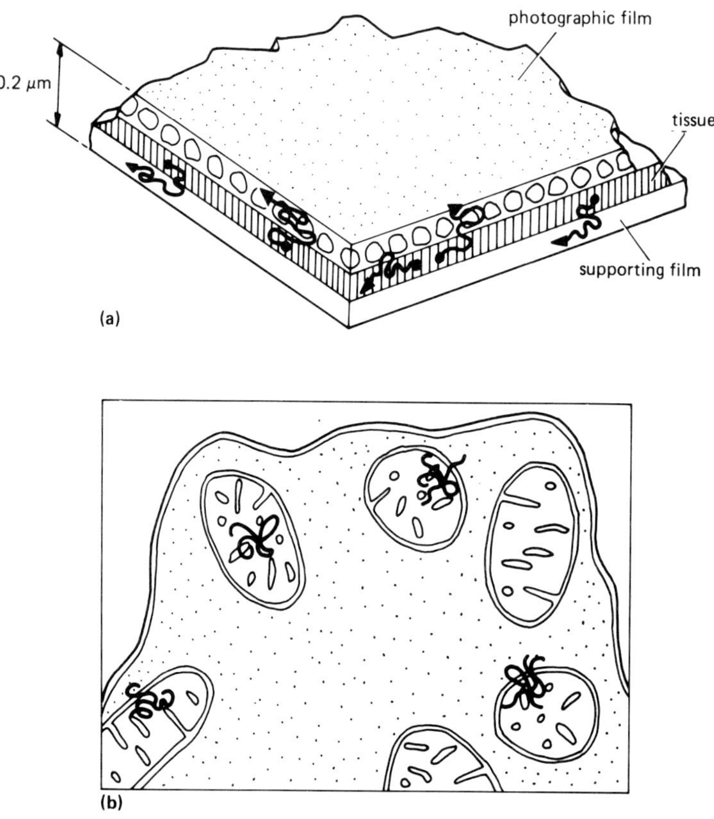

Figure 18 Autoradiography on an electron microscope section. **(a)** The bottom layer of the sandwich is a supporting film, often collodion. The middle layer is the sectioned tissue including tritium or other radioactive atoms. Wiggly arrows show the paths taken by emitted electrons. A photographic film containing a single layer of silver bromide crystals forms the top layer. **(b)** The appearance of an autoradiograph showing a tritium-labelled compound located only in mitochondria. Each group of dark squiggles is made of streaks of silver in the photographic film, marking the course of electrons.

Later, though, more compelling evidence emerged. One experiment involved treating cells with a compound known to be incorporated into DNA. The compound was labelled with tritium or ^{3}H, an isotope of hydrogen. If the cells were treated while the nuclear DNA was being made, most of the tritium could be expected to end up in the nucleus. But if the cells were treated when nuclear DNA was not being made, various other fates could await the tritium. If it was not built into DNA it might be distributed randomly through the cytoplasm. But if DNA was being made somewhere, the label should accumulate at that place.

To use autoradiography in this case, the tissue must first be cut into thin sections. The way this is done is described in chapter 7 (p. 81). If necessary, the sections are supported on a layer of plastic. Then a thin film of photographic emulsion is laid on top of the section (figure 18a). The tritium atoms give off electrons which affect the crystals of silver bromide in the same way that light does. The specimen and the film must be left for several weeks or months to give time for enough electrons to be emitted.

When the film is developed, each crystal hit by an electron turns into a streak of silver metal. Then the whole sandwich is put into the electron microscope. Figure 18b shows the kind of result. As is explained in chapter 8 (p. 91), the silver shows up very dark on the electronmicrograph. Clearly, the labelled molecules have arrived in the mitochondria, suggesting that indeed DNA is formed there. The silver streaks may look like the long-chain molecules of DNA, but in fact each one is merely due to the wiggly route the electron took through the photographic film.

Geiger counter

Autoradiographs are very useful for qualitative observations, such as locating the position of an atom, but sometimes quantitative observations are needed. For instance, the photographic technique just described might not have worked satisfactorily. In that case the label could be located by separating out the different organelles of the cells. This can be done by homogenising the tissue into fractions (see p. 71), and then measuring the radiation given off by each fraction. How can this measuring be done?

The commonest instrument for this is the Geiger–Muller or **Geiger counter**. It is a versatile instrument since it detects anything that will ionise a gas molecule. It is obviously sensitive to electrons or beta rays, but it also responds to other kinds of ionising radiation such as gamma rays and X-rays.

The gas to be ionised is contained in an air-tight metal tube (figure 19). One end of the tube is closed by a thin 'window', often made of mica.

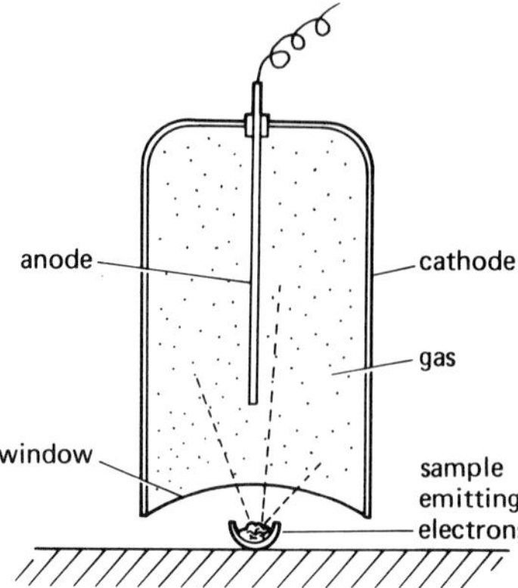

Figure 19 Geiger counter: radiation from the sample causes ionisation in the gas, which in turn generates a current in the external circuit.

The counter is placed over the sample so that some of the radiation can pass through the window into the gas. If beta rays are being emitted from an isotope, each electron will collide at high speed with a neutral gas molecule, knock one of its electrons off, and so convert the molecule into an ion.

The metal wall of the tube and rod down the middle function as electrodes. The rod is the anode and has a strong positive charge. The tube is the cathode and has a strong negative charge. So, as soon as a gas molecule is ionised, the electron knocked off it is attracted to the rod, while the ion is accelerated outwards towards the wall of the tube. But that is not all, for the ion is very likely to collide with other molecules, ionising some of them. In this way, a whole cloud of ions quickly builds up.

Just as in the mass spectrometer, when the positive ions reach the cathode they accept electrons from it. At the same time, electrons are arriving at the anode and being absorbed into it. This sets up a current in the external circuit. Each beta particle (electron) entering the counter creates an ion cloud which in turn produces a pulse of current. So, counting the pulses gives an accurate measure of the number of radioactive particles entering the counter.

At least, the instrument is accurate when not too many particles are arriving each second. However, when they are coming thick and fast the instrument may be inaccurate. This is because, while there is a cloud of ions in the tube, the counter will not respond to any new particles entering the tube. This 'dead time' may last 100 microseconds (μs) or more. So, when the radiation is intense, the Geiger counter underestimates it.

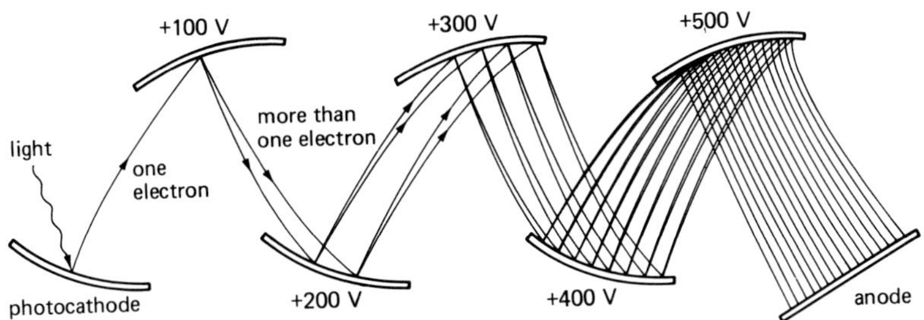

Figure 20 Photomultiplier: light striking the photocathode makes it emit electrons. They pass along the line of charged plates, each time increasing in number, until they are collected by the anode.

Scintillation counter

For some purposes, a **scintillation counter** is preferable to a Geiger counter. It has a much shorter dead time, making it more accurate. It can also be made more sensitive, responding to weaker radiation.

The most important part of the instrument is a substance called the phosphor. When it absorbs ionising radiation, it emits minute flashes of light. The problem then is to measure the scintillations or flashes of light. They are much too weak to see with the naked eye, so each flash must be magnified.

This is done by using a **photomultiplier** (figure 20). A series of metal plates are given different electric charges, increasingly positive, so that electrons are shot off one and immediately attracted to the next in the line. First the flashes of light from the phosphor hit the cathode, called a photocathode because each flash of light causes it to give off one electron. This hits the next plate which responds by emitting more than one electron. The process continues along the line, and each time the number of electrons is multiplied. By the time it reaches the anode, the cloud of electrons is so intense that a detectable pulse of current is produced.

Radiocarbon dating

Any unstable isotope decays at a fixed rate called its **half-life**. In the case of ^{14}C, this means that any sample will lose half its ^{14}C atoms by decay in 5700 years. After another 5700 years, half of the remaining atoms will also have turned into nitrogen; and so it goes on (figure 16b). This makes possible the method of radioactive carbon dating. The amounts of ^{14}C and ordinary ^{12}C are measured in an object such as a recent fossil or an archaeological find. Assuming we know the proportions of the two kinds of carbon in the object originally, it is easy to calculate how many thousand years of decay has produced the change.

At one time, difficulties arose from incorrect assumptions about the original amount of ^{14}C in objects being dated. However, the method has now been made much more accurate by calibrating it against dates determined from counts of annual rings in ancient trees.

This method of radiocarbon dating has made it possible to fit a detailed time-scale to the ecological changes occurring in Britain since the last ice age. Some peat bogs have been accumulating since the ice retreated. Pieces of wood or other plant material from each layer of peat can be dated to show how old the layer is. At the same time, the pollen preserved in each layer indicates what kind of vegetation covered the surrounding countryside. So, we can find out when tundra gave way to birch or pine, and when these were succeeded by oak forest.

5 Chromatography and electrophoresis

We know what photosynthesis does to the oxygen in water. Tracers helped solve that problem. But what happens to the carbon dioxide? The answer to this is now known in detail, thanks to a combination of autoradiography and chromatography.

Many photosynthetic plants capture carbon dioxide by means of the Calvin cycle. This is a complex series of reactions in which carbon dioxide gas becomes part of carbohydrate molecules. Some of Melvin Calvin's experiments are well known. One particular problem was to find which compound the carbon dioxide molecules are first attached to. Calvin's approach was to provide a culture of unicellular algae with ^{14}C-labelled carbon dioxide. Sooner or later a very large number of compounds in the algal cells could be expected to acquire the label, perhaps every single one. On the other hand, if the algae were killed soon enough, only one compound should have the label. How could it be identified?

It would have been easy if there were some way of attracting the labelled molecules to one place, like using a magnet to pick nails out of sawdust. This is not actually such a silly idea: magnets have been used to separate one kind of cell from others, after the former have been persuaded to take up small bits of metal. But this approach will not help us to find labelled carbon atoms. Instead, all the likely compounds must be separated, and each one checked to see if it has the label. This is where paper chromatography has proved so useful.

Paper chromatography

When the algae had been exposed to labelled carbon dioxide for only a few seconds or minutes, they were suddenly killed. The cells were then homogenised, a process described in the next chapter (p. 71), and the soluble compounds extracted.

The principle of any kind of **chromatography** is something like the time-of-flight mass spectrometer (p. 37). A mixture of different molecules or atoms is concentrated at one spot, then they are made to move away in the same direction. They become separated because they move on average at different speeds, not because they go off in different directions as in the magnetic-field mass spectrometer.

So, in the photosynthesis experiment, the mixture of chemicals is placed at the starting position, usually in one corner of a piece of paper. This is done by putting a drop of the solution on the paper using a pin or fine pipette, and allowing it to dry. Then another can be put on the same place and dried. Repeating this many times gives a concentrated spot of the mixture.

Now the paper is suspended with one edge in a dish of solvent (figure 21a–c). The solvent rises up the paper by capillary action. As the solvent front sweeps past the dried spot of mixed compounds, these start moving with the flowing solvent. But they move at different rates. For instance, the amino acid alanine might move twice as fast as glucose phosphate.

This would be enough to separate those two compounds completely. But in a mixture of many compounds, some might travel at the same speed. These would land up in the same small area of paper and so would not have been separated. Take, for example, a solution of glucose phosphate, citric acid and alanine. If a different solvent had been used, it would have been possible to separate the alanine and glucose phosphate, but then the alanine and the citric acid might have finished up on top of each other.

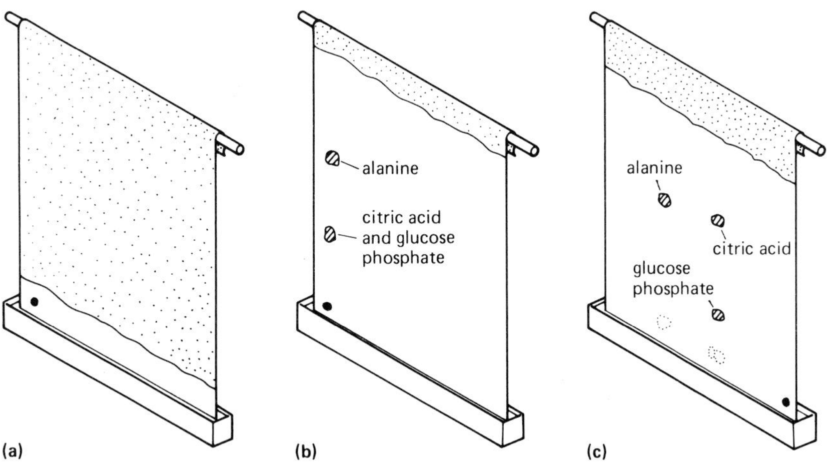

Figure 21 Paper chromatography. **(a)** The paper is suspended with its lower edge in a bath of solvent. A spot of the mixture has been applied at the origin. The solvent front has risen just above the spot. **(b)** Now the solvent front has almost reached the top of the paper, but citric acid and glucose phosphate have not been resolved. **(c)** After turning the paper anticlockwise, the origin has been moved to the lower right. The spots in **(b)** are shown dotted. A second solvent has separated citric acid and glucose phosphate.

The clever thing about paper chromatography is that it can be carried out in two dimensions, length and breadth, and each can be used for a different separation. First one solvent is used, and that separates alanine from the other two compounds. Then the paper is turned through a right angle and suspended with its lower edge in a different solvent. Now the spots move in a different direction, and citric acid and glucose phosphate

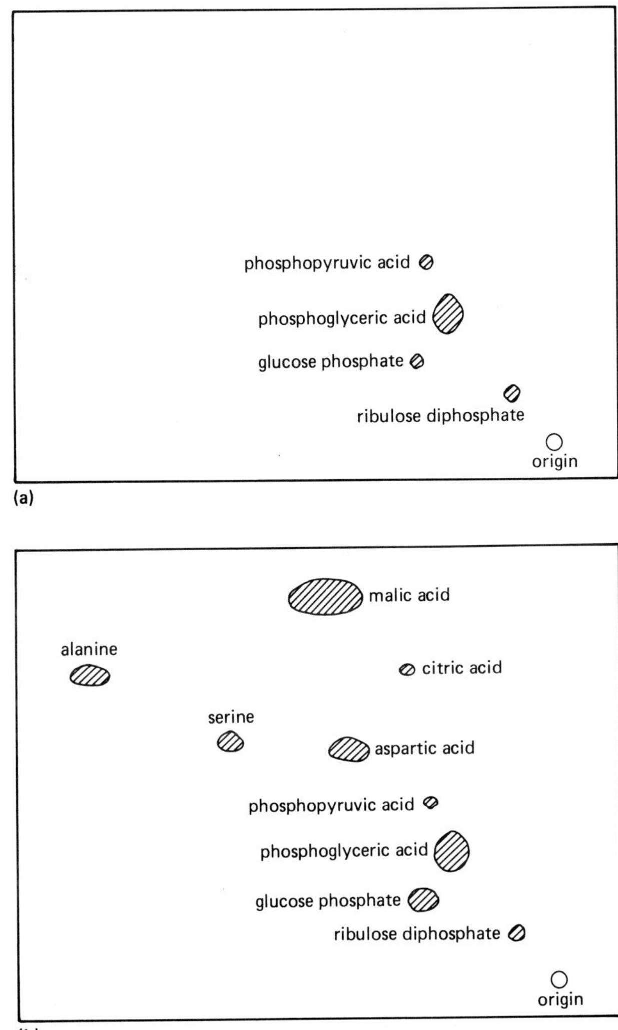

Figure 22 Two autoradiographs made from chromatograms. The substances separated were extracted from algae given $^{14}CO_2$. In **(a)** photosynthesis only continued for 10 s, while in **(b)** the algae were killed after 2 min.

are separated. This trick gives the same effect as if the compounds had moved away from the origin in different directions.

Figure 22 shows two chromatograms from the photosynthesis experiment. In one case the algae were killed only ten seconds(s) after receiving the label, and in the other they were left for two minutes (min). To the naked eye the patches on the chromatogram are invisible, but making an autoradiograph shows up the patches containing the label. At least a dozen compounds are obvious after 2 min, but after 10 s only a few are strongly labelled. In particular, phosphoglyceric acid contains most of it. Thus it was shown that phosphoglyceric acid (PGA) is the first stable compound formed from the carbon dioxide. Note also the positions of the compounds mentioned in figure 21c.

Principles of partition chromatography

Although chromatography is useful, there is still some mystery about how it works. In paper chromatography, the paper itself seems to function merely as a support or framework; it takes no direct part in the separation of the mixture.

The method works because there are two liquid phases, one stationary and one mobile (figure 23). Paper always contains a certain amount of water, in and around the cellulose molecules: this layer of water does not move, and so is called the **stationary phase**. In contrast, the solvent moves and this represents the **mobile phase.**

The speed at which the molecules of the mixture move depends on how soluble they are in the two liquids. Suppose one substance is very soluble in the mobile phase and completely insoluble in the stationary phase. All its molecules will be in the mobile phase, and so it will move up the paper at the same speed as the solvent. In fact, it will keep up with the solvent front. This can be expressed in a quantity called R_F. For any given solute and solvent,

$$R_F = \frac{\text{distance moved by solute}}{\text{distance moved by solvent front}}$$

In the case just mentioned, $R_F = 1$.

At the opposite extreme, a substance which is entirely insoluble in the mobile phase would not move up the paper at all. It would be stuck in the stationary phase, and its R_F would be zero. Most solutes in fact have intermediate values of R_F. In figure 21b the value for alanine is about 0.7 and that of the other two solutes is about 0.3. Separation only occurs where the R_F values are different.

So what is happening when the R_F is between 0 and 1? It must be imagined that each solute molecule moves frequently from one phase to the other, and back again. If this happens often enough, the amount of time each molecule spends in each phase can be calculated. In figure 23, for instance, the concentration in the mobile phase is greater than that in the stationary phase. Each solute molecule spends about twice as much time in the moving solvent as in the stationary water. Like alanine, its R_F is about 0.7, and it moves at two thirds the speed of the solvent. The word 'partition' refers to how the solute molecules divide their time between the two liquids. This method of separating substances is called **partition chromatography**.

Although chromatography superficially resembles the time-of-flight mass spectrometer, it is now obvious that they are really very different. The ions in the spectrometer are like cross-country runners, the faster ones getting ahead of the slower ones. But in chromatography, all the travelling solute molecules are moving at the same speed. That of course is the speed of the solvent. It is rather like an escalator moving upwards beside a stationary flight of steps. The alanine molecules are people who spend two thirds of their time on the escalator, and for the remaining third they are

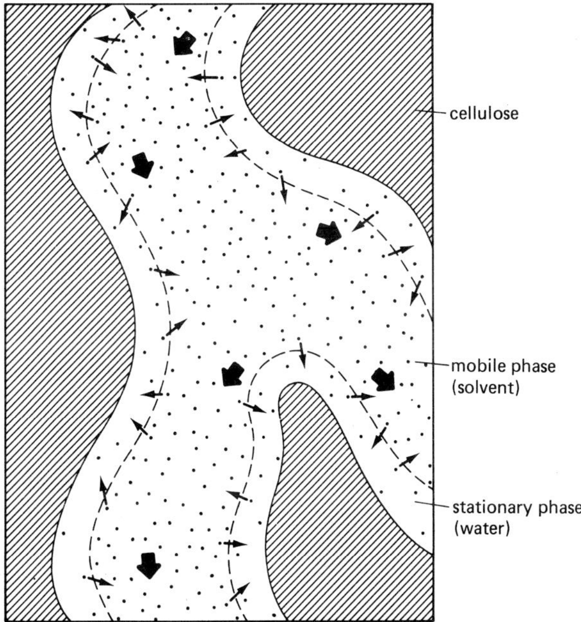

Figure 23 Partition chromatography. The stationary phase is water attached to the cellulose fibres. Large arrows indicate the flow of the mobile phase. Solute molecules are shown as dots, and small arrows represent molecules changing from one phase to the other by colliding with the boundary.

resting or admiring the view at various places on the steps. The glucose phosphate molecules are like people who spend only a third of the time on the escalator.

Resolution

Inevitably the size of a spot gets larger as it travels up a chromatogram. This is not important when two spots are well separated, like alanine and glucose phosphate in figure 21b. However, the enlargement of spots is more serious with the citric acid and glucose phosphate. If the spots were smaller they might have been separated, but because they have expanded they overlap.

When we speak of the **resolution** of a chromatogram we mean its efficiency in separating two compounds (figure 24). This is really the same as sensitivity in an instrument. A suitable choice of solvent may make the R_F values very different, and this would give good resolution. Alternatively, separation may be achieved by keeping the spots small, so that spots close together do not overlap. How can this be done? Well, obviously, it helps to start off with as small a spot as possible at the origin. This is why a pin or extremely fine pipette is used. After that it is necessary to encourage the solute molecules to change frequently from one liquid phase to the other. If they do not change phases frequently, the molecules may spend different proportions of their time in the mobile phase. Therefore, they will have different average speeds, and this will cause them to spread out along the paper.

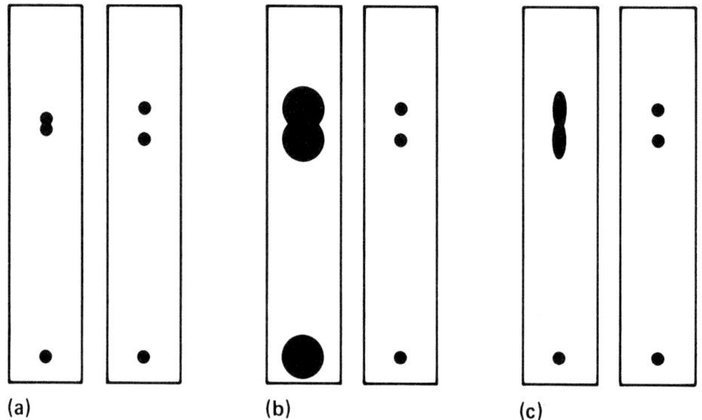

Figure 24 Resolution. In each of the three cases, resolution is poorer in the left-hand diagram than in the right-hand one. The origin is the spot at the bottom. Only one solvent has been used to separate the mixture in one direction. Poor resolution is due to: **(a)** R_F values being too similar; **(b)** the origin spot being too large; **(c)** solute molecules not changing often enough between phases.

How can the solute molecules be encouraged to cross the boundary between the phases? There are two possibilities. First, the solvent front can be made to move slowly. This will allow more time for solute molecules to collide with a boundary and so enter the other phase. However, if the solvent front moves *too* slowly simple diffusion will make the spots expand, so that any advantage in resolution will be lost. Second, the area of the boundary can be increased. This will make collisions more frequent. The area of the boundary is determined by the surface area of the cellulose, so using paper with thinner fibres can improve the resolution.

Column chromatography

Paper chromatography is excellent for making qualitative observations. For example, in the photosynthesis experiment, all that is necessary is to tell which compounds have the label. But what can be done if we want to measure the amount of label in each substance?

One approach would be to cut out each spot from the chromatogram with scissors. Then its radioactivity could be measured with a Geiger or scintillation counter (see pp. 43–45).

A more usual approach is to use **column chromatography**. Here the aim is to separate out all the components so that the quantity of each can be measured. First, a solid material in finely divided form is packed into a vertical tube (figure 25a). Some of the materials used are sugar, calcium carbonate and magnesium oxide. Then a solution of the substances to be separated is carefully added to the top of the column. When it has just soaked in, more solvent is added and kept continually topped up. The solvent, acting as the mobile phase, flows down the column, and the components of the mixture separate out as in paper chromatography. If the position of each zone is known the solid filling can be pushed out of the bottom of the tube with a plunger, and each zone cut off with a knife. The separated compounds can then be washed off the solid with a solvent.

Another method involves collecting the drips from the bottom of the column (figure 25f). Machines are available which can count the drips, so that when a set number have fallen into one test tube another tube is automatically moved under the column. In this way the separated mixture is collected as 100 or more samples. Each component of the mixture may occupy several of the tubes. The machine is known as a **fraction collector**, because each sample is referred to as a fraction.

If the concentration of solute in each fraction is measured, a graph can be drawn (figure 25g). Each zone or separated component shows up in the graph as one peak. If all the fractions have the same volume, the total quantity of each component is given by the area under its peak.

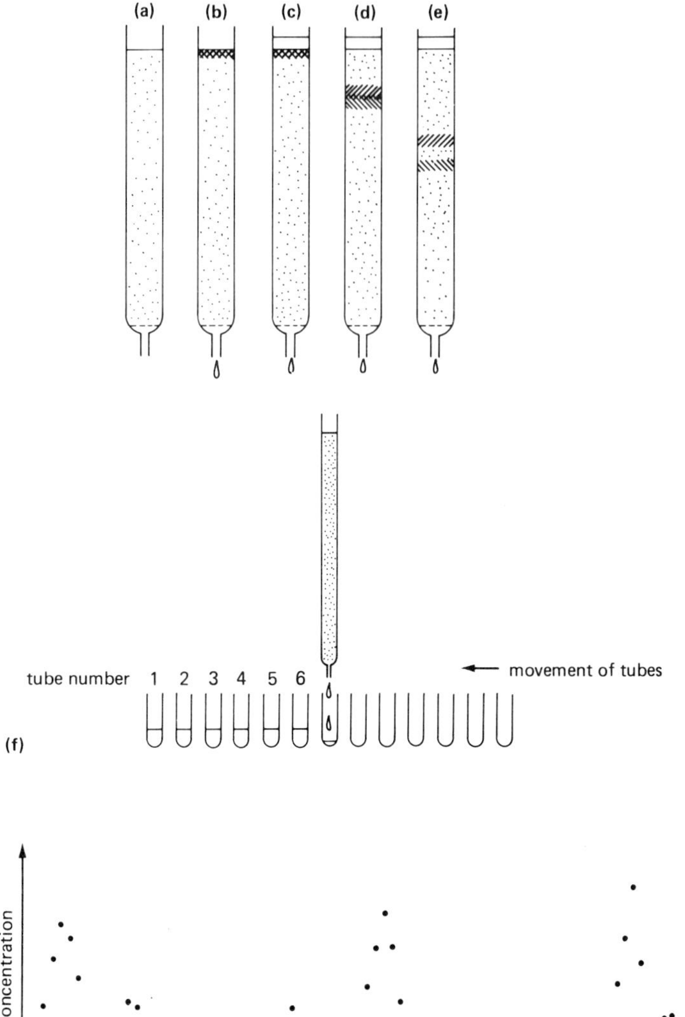

Figure 25 Column chromatography. In **(a)** the column is packed with a solid material, and in **(b)** the mixture has been allowed to soak into the top few mm. Solvent is added **(c)**, and as it flows down the column the mixture moves down and separates **(d)** and **(e)**. **(f)** A fraction collector automatically collects a fixed volume of solvent from the column before the next tube is moved into place. **(g)** Concentration of solute in each tube is plotted against tube number. The area under each peak of the graph gives the total quantity of the component.

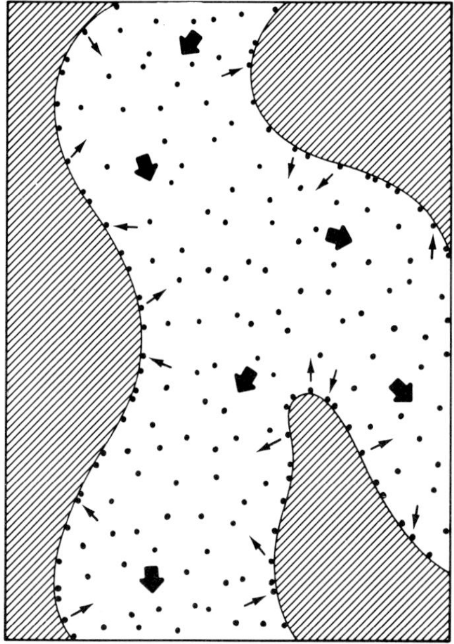

Figure 26 Adsorption chromatography. Adsorbed solute molecules adhere to the solid surface.

The figure shows the kind of result expected when the amino acids are separated after a protein has been broken down.

Some column chromatograms may work like paper chromatography. In other words, the solid may just be supporting a stationary liquid phase, in which case it is an example of partition chromatography (figure 23). But in other cases, the mechanism is slightly different. The stationary phase may be the solid itself. Instead of dissolving in a stationary liquid, the solute molecules spend part of the time adsorbed on the solid. In adsorption, the molecules stick to the solid surface, but not because of the attraction of unlike charges. Otherwise, the principle of this **adsorption chromatography** (figure 26) is very like partition chromatography.

Finding the zones

It was hinted earlier that finding the components of a mixture after separation may be difficult if they are invisible. In the photosynthesis experiment an autoradiograph solved the problem. In a fraction collector, the positions of the amino acids may not be known until the concentration of solute in each test tube is measured.

A variety of other methods is available. Particularly useful is ultraviolet absorption. Many compounds, such as proteins and nucleic acids, absorb ultraviolet radiation of 254 nm wavelength. Inspection of a paper chromatogram under ultraviolet light will therefore reveal the spots as dark patches. Some compounds fluoresce under ultraviolet light, so these will be seen as bright patches.

Yet another approach is to stain the paper chromatogram or the fractions. For instance, adding ninhydrin to amino acids makes them go brown. Some compounds are already coloured, for example plant pigments. The first use of chromatography was by the Russian botanist Mikhail Tswett soon after 1900; he separated photosynthetic pigments in a column of calcium carbonate. This is where the word 'chromatography' comes from, since it means 'writing with colours'. However, Tswett's method only became widely used in the 1930s.

The problem of proteins

Proteins are one of the most important groups of compounds, but separating them proved difficult with most methods. In the early days, blood plasma proteins had to be classified by their solubility in salt solutions. A certain concentration of ammonium or sodium sulphate was used; the proteins which were precipitated were called **globulins**, and those left in solution were called **albumins**. If another strength of salt solution was used, some of the globulins, called fibrinogen, could be separated from the other globulins. One or two other categories could be distinguished, but it later became obvious that each of these 'kinds' of protein was really a mixture of many different compounds.

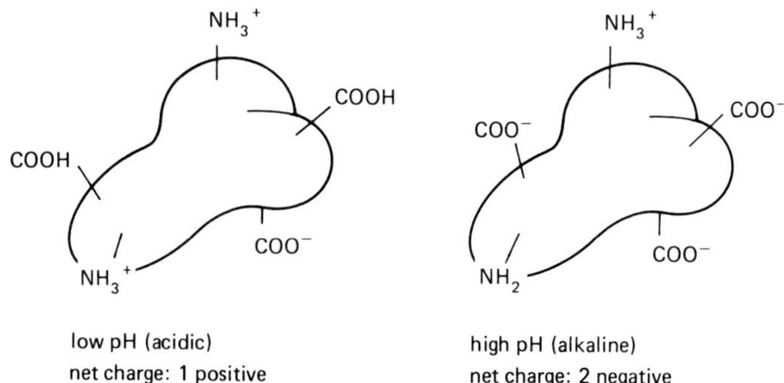

Figure 27 Proteins. A protein molecule bearing on its surface five groups which can be ionised. The degree of ionisation depends on the pH of the surroundings.

One particular type of chromatography has proved useful with proteins. It depends on the fact that proteins are charged molecules. Each one carries a net positive or negative charge. This is not because there is just a single atom on each protein which is ionised; on the contrary, a protein may be bristling with them. An amino group on the surface may be ionised as $-NH_3^+$, so carrying a positive charge. Strictly speaking, it acquires an extra proton or hydrogen ion, which gives it the positive charge. Somewhere else, a carboxyl group may be ionised as $-COO^-$, carrying a negative charge. The overall charge on the molecule is determined by the difference between the numbers of negative and positive charges (figure 27).

Another important feature of proteins is that this overall net charge can change. When the surrounding solution is fairly acidic, with a low pH, the amino groups will tend to be ionised and the carboxyl groups not, so the net charge might be positive. In an alkaline solution, with a high pH, the reverse is more likely, giving a negative charge. This means that, for separating proteins by chromatography, the pH must be very carefully controlled by using buffers. A buffer is a substance which cuts down any fluctuations in the pH.

Ion-exchange chromatography

The kind of chromatography found to be the most effective with proteins depends on **ion exchange**. Various gels, modified celluloses, or resins are suitable as the stationary phase. What they all have in common is charged groups on the surface. If these are negative, then positive ions passing in the mobile phase will be attracted and will spend some time stuck to the surface.

When the column is set up, the solid material is already loaded with hydrogen or some other positive ion. Then as the solution of mixed proteins is added, there is an exchange of ions. Some inorganic ions move from the solid surface to the solution, and some positively charged proteins adhere to the surface by the attraction of unlike charges (figure 28). There is a continual movement of molecules back and forth between the surface and mobile phase. In this way it is just like adsorption chromatography. Some proteins spend on average longer on the surface than other proteins, because they have a stronger positive charge. This means that these move down the column more slowly, and so different proteins can be separated. The method only works if the pH is carefully regulated, since a change of pH alters the charge on the proteins.

Ion-exchange chromatography has also been important in separating nucleotides, the building blocks of nucleic acids. DNA or RNA can be broken down to the four constituent nucleotides, and the exact quantity of

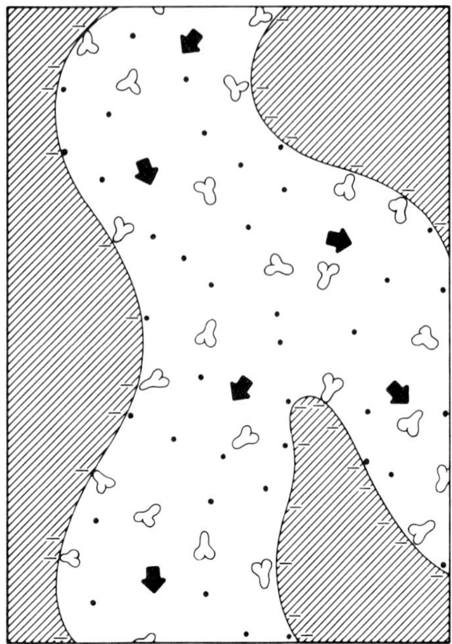

Figure 28 Ion-exchange chromatography separating proteins. The minus signs show where there are negatively charged places on the solid stationary phase. Some attract hydrogen or other small ions, shown as dots, and others attract protein molecules which are the large particles.

each one measured. In this way Chargaff found that in DNA the ratio of adenine and thymine molecules is 1 : 1, as is the ratio of guanine and cytosine. This was one of the most important pieces of evidence supporting the double-helix theory of DNA.

Electrophoresis

The problem of separating proteins has been tackled using another method. This also depends on the fact that proteins have different electric charges. You will remember that mass spectrometers cause ions to shoot along a tube, attracted by the strong electric charge on an electrode. The same kind of thing can be done with proteins in solution, although the rate of movement is very much slower. This is the principle behind **electrophoresis**.

Electrophoresis requires a positively charged electrode, another one negatively charged, and a solution between them capable of conducting electricity (figure 29a). The solution can be held in a strip of filter paper. When the mixture of proteins has been spotted onto the paper, just as in

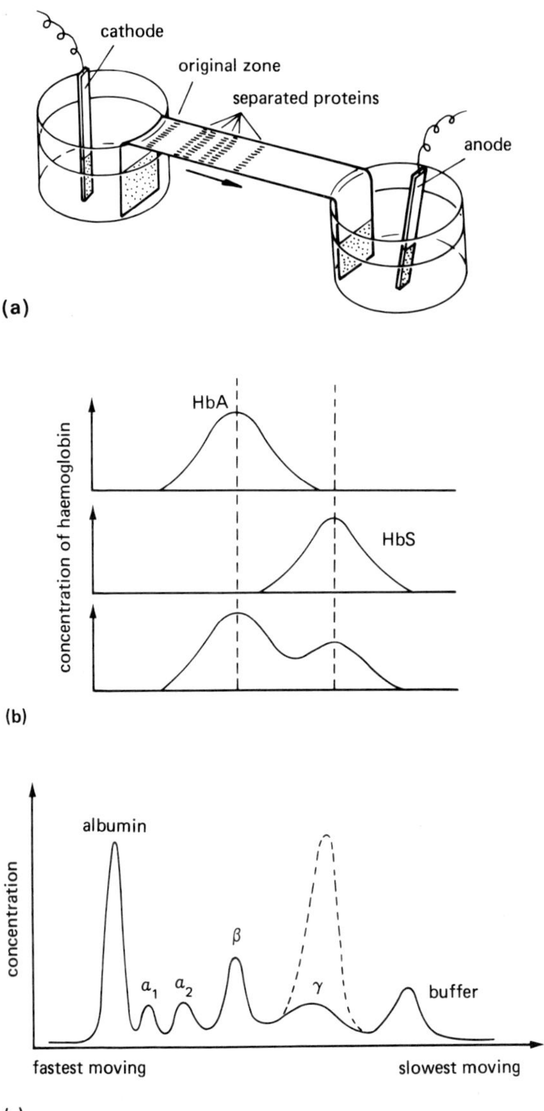

Figure 29 Electrophoresis. **(a)** In zone electrophoresis, a filter-paper strip soaked in a solution connects two dishes containing the solution and electrodes. In this example, all the proteins are negatively charged and so are applied to the paper near the cathode. They move at different rates towards the anode, separating into zones. **(b)** Analysis by zone electrophoresis of the haemoglobin from a normal person (*top*), someone suffering from sickle-cell anaemia (*middle*), and a heterozygote with sickle-cell trait (*bottom*). **(c)** Analysis by moving-boundary electrophoresis of serum proteins. Albumins move fastest, followed by the globulins: α_1, α_2, β and γ. The dotted line shows the enhanced amount of γ-globulins in certain diseases.

paper chromatography, the circuit is switched on so that the electrodes become charged. Then the proteins migrate towards the electrode of opposite charge. Positively charged proteins move towards the negative cathode, and the more strongly charged they are the faster they move. Negatively charged proteins migrate in the opposite direction. Buffers must be used to prevent any change in the proteins' charges.

If haemoglobin is extracted from red blood cells, it can be analysed by electrophoresis. When taken from a normal person, it will form a single zone, indicating that it is just one kind of protein. The same procedure using haemoglobin from someone with sickle-cell anaemia also gives just one zone, but in a slightly different place. This suggests that the protein is different; it is called haemoglobin S to distinguish it from the normal haemoglobin A (figure 29b). It is now known that there is only one difference in the amino acid sequence in the two proteins; the sixth amino acid from one end is glutamic acid in haemoglobin A but valine in haemoglobin S. Glutamic acid carries a negative charge whereas valine does not. Thus the two proteins have different charges, and this explains why they can be separated by electrophoresis.

The normal person and anaemic person are both homozygous for the gene coding for the protein, so what kind of protein does the heterozygote have? It could be a third kind of molecule, or instead a mixture of A and S. Electrophoresis quickly answers the question. A graph showing concentration of protein in the zones is bimodal; it indicates that both proteins are present, though S is slightly less abundant than A.

Electrophoresis has also proved useful in improving the classification of blood proteins. Not only can albumin and globulin be distinguished, but it is apparent that there are several kinds of globulin. The usual procedure is called moving-boundary electrophoresis, but it gives essentially the same result as zone electrophoresis. Moving boundaries are encountered again in the next chapter (see p. 69). Figure 29c shows the result of separating serum proteins by electrophoresis. The fastest moving proteins, the albumins, are on the left. Then follow at least four kinds of globulin.

This kind of analysis can be useful in medical diagnosis. Most of the antibodies produced in the immune response are gamma globulins. If the gamma globulin peak is much larger than normal, it indicates an infection or other disorder in the body. The effect is marked in multiple sclerosis.

Ouchterlony's method

Separating the serum proteins into five groups is nowhere near the goal of isolating pure proteins. A step nearer is achieved by Ouchterlony's method. This exploits antigen–antibody reactions. Antigens are usually proteins or polysaccharides, and an antigen will form a precipitate only with one particular antibody protein. An antigen can therefore be used to pick out that one protein from a mixture.

In practice a layer of agar is prepared, and two wells are cut in it. This used to be done in a petri dish, but now it is more often performed on a microscope slide. A mixture of antibodies is placed in one well, A, and antigens in the other, B (figure 30). All the proteins diffuse outwards through the agar. When an antigen meets an antibody that reacts with it, they precipitate in the agar. This is seen as a thin line. Because proteins diffuse through the agar at different rates, these lines of precipitate form in different places, one line for each antigen–antibody pair.

But this in itself will not identify the proteins in the two mixtures, or even in the lines of precipitate. To achieve this, a third well is cut in the agar. Into this can be put a pure protein, either an antibody or an antigen. Suppose we are studying diphtheria toxin, and need to know whether it is present in the mixture in well B. It is arranged that the mixture in well A has the antibody to the toxin, and the pure toxin is put into well C, so a line of toxin–antibody precipitate will form between A and C. Now, if none of the lines between A and B is formed by the same two proteins, the pattern will be like figure 30a. The lines will cross. But if one of the lines between A and B is caused by the same reaction, it will not cross the toxin–antibody line. Instead it will join it, forming an obtuse arrow head (figure 30b). This means that the toxin has been identified as present in the

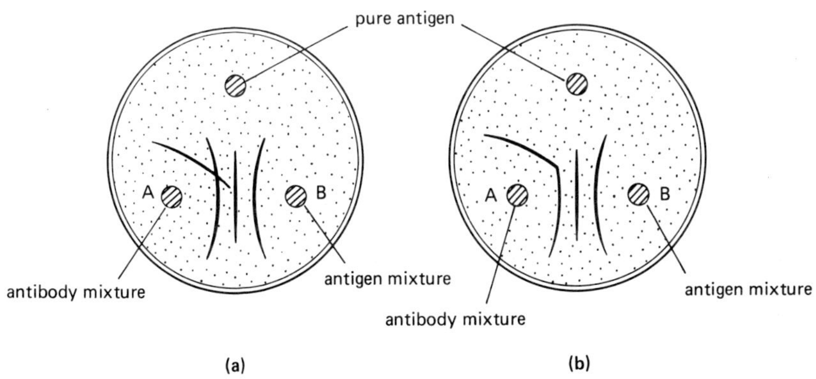

Figure 30 Ouchterlony's method. In each case the wells A and B cut in the agar plates contain unknown mixtures of proteins. Each antigen–antibody pair causes a line of precipitate. The pure antigen in the third well is used to identify the reactions between A and B. In **(b)** the pure antigen is also present in B, but in **(a)** it is not.

mixture in B. It is not easy to measure the amount, but Ouchterlony's method allows valuable qualitative observations to be made.

When pure proteins were difficult to prepare, this method was limited in its use. But pure proteins called **monoclonal antibodies** are now much easier to produce in large quantities. They are made by lymphocytes, and each lymphocyte produces only one antibody. The trick is to isolate a single lymphocyte and fuse it with a tumour cell. This causes the hybrid cell to multiply many times, giving a clone of genetically identical cells manufacturing the pure protein.

In this chapter, several methods for separating compounds have been mentioned, but this is only scratching the surface. All these methods have been combined or developed in ingenious ways, so that now the most unlikely separations can be performed. For instance, **immuno-electrophoresis** involves separating proteins by electrophoresis and then precipitating them with antibodies, all on the same agar plate. It is even possible now to tip a mixture of cells into a chromatography column and separate the different kinds of cells. A good account of these methods is given by Williams and Wilson in their *Principles and Techniques of Practical Biochemistry*. The next chapter describes how the centrifuge can also be used for separating things and measuring them.

6 Centrifuging

What instrument can measure relative molecular mass, the density and approximate shape of large molecules, and isolate DNA or mitochondria from cells? The answer is, the centrifuge. For many investigations, it is the biologist's best friend.

In practice, different models are needed for different tasks. One set of tasks involves separating and purifying a chosen kind of particle or substance. It might be preparing a suspension of mitochondria for further study. For this a **preparative centrifuge** is required. This is the kind most students are familiar with. In research, however, another model is frequently used, called an **analytic centrifuge**. The purpose of the analytic centrifuge is not to get something ready for a later experiment, but actually to make detailed measurements while the centrifuge is spinning. It is possible to describe only one or two of these measurements in this chapter.

Sedimentation by gravity

The principles of centrifuging can be explained by an experiment which almost everyone will have tried. If some soil is vigorously shaken up with water in a jam jar, all the particles become dispersed evenly through the water (figure 31). As soon as the jar is put down, the sand falls to the bottom. Then a layer of silt particles settles out on top of the sand. Finally, perhaps after several days, the water becomes fairly clear once more, and very fine clay particles are seen to have formed a third layer of sediment.

An identical result would have been achieved if the soil suspension had been spun round fast in a centrifuge. In the jam jar, the process is slower because the force causing the particles to settle out is only the weak force of gravity. But, simple though it is, this experiment poses four questions which we can ask about centrifuging.

First, are the layers of sediment pure? In general, the answer is 'no'. Take the sand layer, for example. While some sand grains were falling through the whole column of water, other silt and clay particles were already very close to the bottom. They had only a short distance to fall, and so could have reached the bottom even before some sand grains. So the sand layer, although it contains all the sand, is also contaminated with silt and clay. For the same reason, all the layers will be contaminated

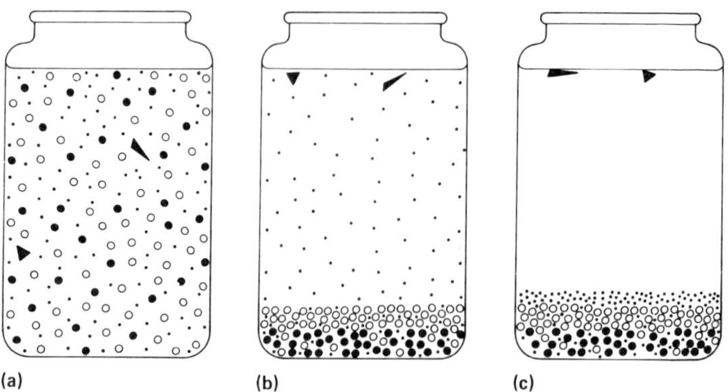

Figure 31 Mixture of soil and water, **(a)** just after mixing, **(b)** during settling, and **(c)** at equilibrium. Solid circles, small sand particles; open circles, silt; dots, clay; triangles, fragments of wood.

except for the topmost; the layer of clay should not contain any silt or sand. Exactly the same applies in centrifuging, so, to get mitochondria reasonably pure, they need to be 'washed' several times. This 'washing' means shaking the mitochondria up in clean liquid, centrifuging to precipitate them as a pellet, and throwing away the liquid which now contains some of the impurities. When this has been done several times, most of the rubbish, such as bits of endoplasmic reticulum, will have been got rid of.

The second question is, what causes the particles to land up where they do? It is probably obvious that in most cases this is density. The sand, silt, and clay are all at the bottom because they are denser than water. There may also be a few fragments of wood floating at the surface of the water; this is because they are less dense than water. The wood and the sediment have been separated very effectively, but this is not true of the three layers of sediment. One way round this might be to pour the suspension into a new jam jar as soon as the sand had settled out, to allow the silt to settle in the new container. When the silt had fallen out of the water the suspension could be poured into a third jar to collect the clay. But as already explained, the sand and silt fractions would be impure. In centrifuging, substances can be separated much more effectively by using a density gradient; this will be explained a little later.

But density differences may not explain everything going on in the soil experiment. In fact, the third question is, how fast do the different particles fall through the liquid? This may have nothing to do with density at all. Sand is made of quartz and clay is not; but suppose in figure 31 that all the particles were made of quartz, so that they all had the same density.

Then the sand-size grains would reach the bottom first, because they fall fastest. The silt-size grains, being smaller, would descend more slowly and so settle on top of the sand. Tiniest of all, the clay-size particles would arrive last to form the topmost layer. So, the order in which particles arrive at the bottom is just as much a result of their size and shape as it is of their density. The larger and more spherical a particle is, the faster it falls. This is because the ratio of its area and mass is small. In this way you could even get a layer of dense particles settling on top of a layer of less dense particles, just because the former were very small and rod-shaped. In fact, one of the most important uses of an analytic centrifuge is to measure how fast particles move through a liquid.

Finally, can all the suspended particles be separated from the liquid? In the soil experiment, some particles may never settle at all. Very small clay particles carrying electric charges would probably never form a sediment, partly because they repel each other electrostatically. Another way in which they remain dispersed is through Brownian motion, the continual buffeting action of other particles colliding with them. Sedimentation by gravity is therefore not often of much practical use, since the force of gravity may not be strong enough to overcome electric forces and Brownian motion. However, centrifugal forces can be made enormously stronger. For this reason, centrifuging can make much smaller particles settle than can gravity.

Centrifuge design

The very simplest centrifuge would be something like figure 32a–b. When at rest, the buckets hang vertically. Test tubes are filled with suspension and one put in each bucket. As the motor speeds up and spins the rotor the buckets swing out to a horizontal position. After a time a sediment is found to have collected in the bottom of each tube.

When the aim is to collect the sediment from a suspension, another kind of rotor is found to be better. This is the **angle-head** design in which the tubes are held at a fixed angle (figure 32c–d). It gives a faster and more thorough sedimentation than the swing-out rotor, as if it were spinning faster than it actually is. The reason for this lies in the resistance particles have to overcome when moving through the liquid. In a swing-out centrifuge, each particle has on average to travel half the length of the tube before it joins the sediment. But in the angle-head, each particle moves only a short distance, horizontally through the liquid, until it meets the outer wall of the tube. After that it slides down the wall to the bottom, and the resistance to its movement in this stage is much less.

These two preparative rotors are very unlike the analytic rotor. In this, there must be a way of observing the suspension, and if possible photo-

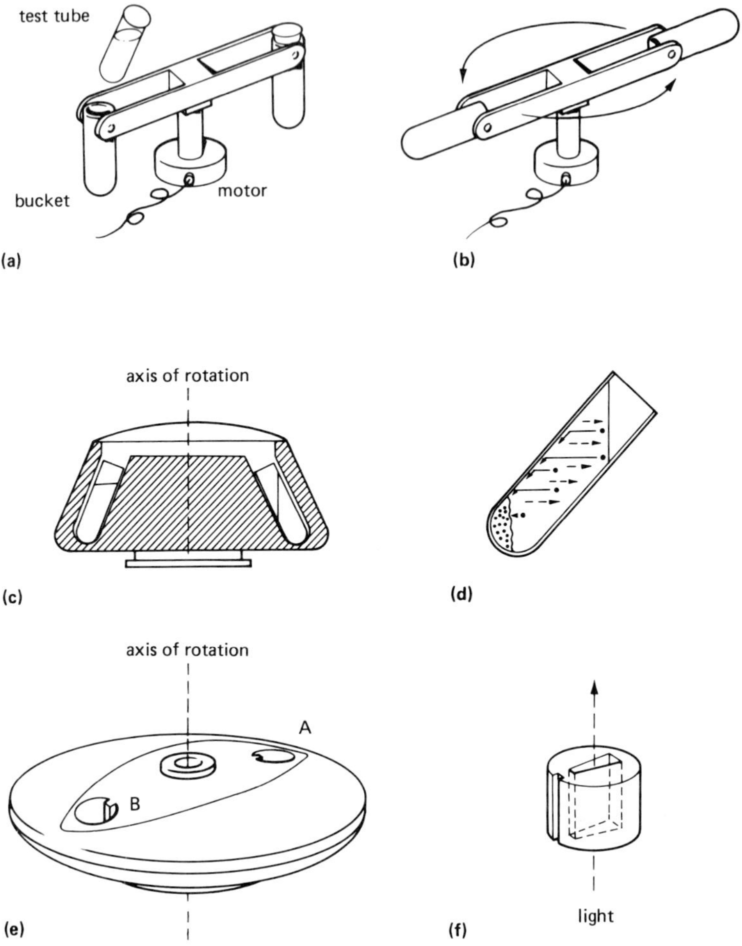

Figure 32 Centrifuges. **(a, b)** A simple centrifuge with swing-out or swinging-bucket rotor; **(a)** at rest, and **(b)** in motion. **(c, d)** Angle-head rotor. In **(c)** the left side shows the meniscus when the rotor is at rest, and on the right when spinning. **(d)** The route taken by sedimenting particles. Interrupted arrows indicate the compensating flow of liquid. **(e, f)** Analytic rotor. In **(e)** the holes A and B hold analytic cells, one of which is shown enlarged **(f)**. It contains a wedge-shaped cavity aligned along the rotor radius. Light can be shone up through the windows forming the top and bottom of the cavity.

graphing it, while the centrifuge is running. Therefore, a special container or 'cell' is used which enables a beam of light to be shone up through it (figure 32e–f). Since the light is often ultraviolet, special windows of quartz or some other material are needed for the top and bottom of the cell. Most of the time, of course, the light is blocked by the rotor, but the analytic cell may pass across the beam of light over 1000 times per second, so that continuous observation of the suspension is possible.

Apart from the rotor design, the main difference between centrifuges is in the speed at which they rotate. The faster the rotation, the larger the **centrifugal force** generated. Some people say there is no such thing as centrifugal force, and in a sense they are right. Suppose the simple centrifuge in figure 32b had a ball bearing in each bucket, and the bottom of one bucket burst. The ball bearing would be thrown away from the centrifuge, but not directly away from it along a radius. It would be thrown off at a tangent, that is, along a line at right angles to the radius (figure 33). This is the direction the ball bearing was moving in when the bucket broke. The ball bearing would have moved from A to A′ if it stayed in the bucket being spun around C, but when the bucket broke it moved from A to B. For a short time after the bucket broke, the points C, A′ and B are still almost in a straight line; and the centrifugal force is said to have made the object move from A′ to B outwards along the radius. So from a commonsense point of view, there is no centrifugal force; but it is a useful shorthand way of explaining why the particles in a suspension sediment outwards, away from the axis of rotation.

Centrifugal force takes the place of gravity in the jam-jar experiment. One difference between the two forces is that centrifugal force can be made as large as we need, just by spinning the rotor faster. General-purpose centrifuges spin up to 6000 revolutions per minute (rpm); these produce a maximum centrifugal force of about 6000*g*, that is, 6000 times the force of gravity. High-speed centrifuges operate up to 25 000 rpm

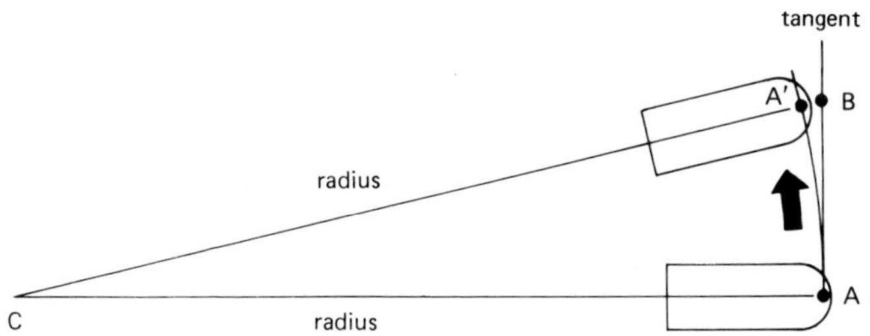

Figure 33 The bucket is rotating anticlockwise about the axis of rotation C. The ball bearing at A moves to A′ if it remains in the bucket. If the bucket breaks, the ball bearing moves along the tangent to B instead of to A′.

and 90 000*g*. These enable the larger organelles from cells to be sedimented, but the centrifugal force is not strong enough to sediment smaller organelles and nucleic acids. Even faster instruments are the ultra-centrifuges, reaching 75 000 rpm and over 500 000*g*; these are used for analytic purposes, and also preparatively for isolating viruses, ribosomes and microsomes.

Density gradients

In their famous experiment on DNA replication (see p. 17), Meselson and Stahl first cultured bacteria in a medium containing ^{15}N, so that the DNA produced was denser than normal. Then the medium was changed back to a solution containing only ^{14}N, so that any new DNA would incorporate the ^{14}N and have a lower density than the old DNA. The problem was to measure the density of the DNA molecules to find out how the new ones were being made.

In a centrifuged suspension, there are two occasions when a particle will not be moved by the centrifugal force. One is when it has reached the end of the tube, and the other is when it is in a liquid of the same density as itself. The trick Meselson and Stahl used was to prepare a **density gradient** in the liquid containing the DNA. If a really strong gradient is needed, a dense solution can be poured into the bottom of the tube, then a slightly less dense one added on top, and so on until the tube is full. This gives a rather uneven change in density from bottom to top of the tube; but it can be made more continuous by gentle stirring. A better result is achieved by pouring two liquids simultaneously into the tube, one less

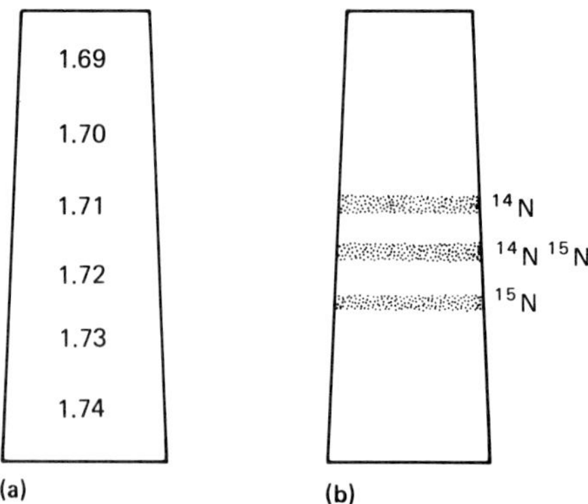

Figure 34 Density gradient. **(a)** An analytic cell seen from above and designed to separate two kinds of DNA. The cavity widens away from the axis of rotation. **(b)** The final positions of ^{15}N DNA, ^{14}N DNA, and DNA containing equal numbers of ^{14}N and ^{15}N atoms.

dense than the other. By gradually increasing the flow of the first and decreasing the second, the density of the mixture can be evenly changed.

The idea of the density gradient is to add the DNA, then centrifuge the tube until the molecules of DNA arrive in liquid of their own density (figure 34a). DNA made with ^{14}N has a density of about 1.710 g cm $^{-3}$, and DNA containing only ^{15}N is 1.725 g cm $^{-3}$, so the gradient needed must include these two densities. The result expected is shown in figure 34b. You will remember that in the jam-jar experiment the particles' size and shape had to be considered as well as their density. But the beauty of the density-gradient method is that size and shape can be ignored. This is because density is the only feature of the DNA which determines where it eventually arrives in the tube.

In fact, Meselson and Stahl did not need to prepare a density gradient by one of the methods already described. They simply filled an analytic cell with the DNA mixed with a solution of caesium chloride having a density of 1.71 g cm $^{-3}$. They then spun this mixture in an analytic ultracentrifuge at 45 000 rpm (140 000*g*) for nearly a day, and this itself generated the density gradient. The very large centrifugal force pushed the heavy caesium atoms away from the axis of rotation. So the density near the axis decreased while that at the outer end of the analytic cell correspondingly increased. The bands of DNA could then be located simply by photographing the analytic cell when ultraviolet light was shone up through it as shown in figure 32f. Meselson and Stahl's original account is well worth looking at. Anyone who can design and carry out an experiment as elegant as this one deserves a Nobel Prize.

S-rates

The method just described involves separating particles according to their density, but biologists often find it more useful to separate them by their sedimentation rates. This is the speed at which the substances move outwards in a spinning centrifuge cell. For instance, there are many kinds of RNA which differ in the length and shape of their molecules. They may all have the same density, so a density gradient would not be of much use in distinguishing between them. But variable size and shape make their sedimentation rates or **s-rates** markedly different.

An analytic centrifuge is needed for this, because the rate of movement of each kind of particle must be measured during the centrifugation. Fortunately, the different kinds of RNA do not need to be separated first, if the **moving-boundary method** is used. Figure 35a shows that the moving boundary of each kind of molecule is originally at the inner end of the analytic cell. As the rotor spins, the molecules move outwards, some kinds faster than others. The rear edge of each kind of molecule makes a

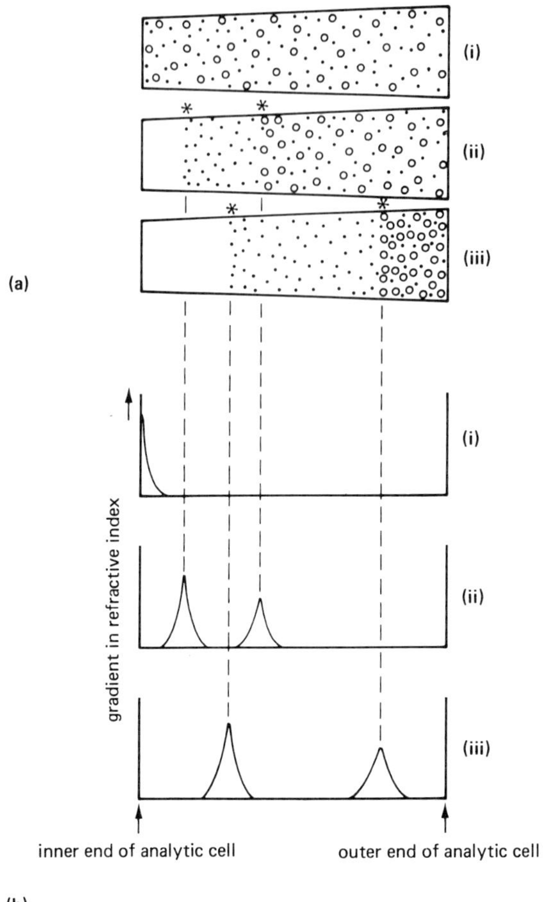

Figure 35 S-rate. **(a)** The moving-boundary method. Asterisks mark the boundaries. In (i) spinning is just starting, and all the boundaries are at the inner end of the analytic cell. (ii) and (iii) are progressively later stages. **(b)** shows how the boundaries are revealed by Schlieren optics.

fairly sharp line or boundary, and it is the speed of this boundary which is measured.

Since the entire contents of the analytic cell may be transparent, not even absorbing ultraviolet light, finding the boundaries might seem an insuperable task. However, where the concentration of solution changes, as it does at each moving boundary, so also does the refractive index (see p. 79). Light rays are bent where the refractive index changes, and this enables the boundaries to be located with an optical arrangement rather like that used to 'see' the patterns of pressure in a wind tunnel. The instrument draws a graph of the change or gradient in refractive index, so that each boundary shows as a peak. All that is necessary is to measure how fast each peak in the graph moves along the analytic cell. The peaks in figure 29c are produced in the same way.

Particles having different s-rates are usually described in terms of the unit S or Svedberg. This is named after Theodor Svedberg, the Swedish scientist who invented the analytic centrifuge in the 1920s. Generally, larger S values indicate larger particles. Numerous kinds of RNA have been distinguished in this way. Small 4S molecules of uniform size are transfer RNA, joined to amino acids before entering protein synthesis on the ribosomes. Larger 18S to 80S molecules comprise messenger RNA; the variable size is due to the different lengths of genes or the polypeptide chains for which they code. A whole range of molecules with various S values can be found in the nucleolus, and it is now known how some of these are assembled in the cytoplasm with proteins to make ribosomes. Without sedimentation rates, research into RNA would not be nearly so advanced.

Homogenisation and fractionation

One of the main uses of the preparative centrifuge is to isolate organelles. But nuclei, ribosomes and similar organelles cannot be separated and purified just by centrifuging cells or lumps of tissue. Cell membranes can be quite tough, and cell walls even more so. Before centrifuging, these must be broken down to release the organelles from the cells, and this is called **homogenisation**.

The methods of homogenising tissues comprise a catalogue of violence beside which TV blood-and-guts programmes pale into insignificance. Particular violence is needed for plant tissues and bacteria. Some homogenisers force the cells through the narrow gap between a piston and a cylinder. The Waring Blender has very sharp spinning blades set at right angles to each other, and these set up powerful shearing forces in the liquid medium; this machine is a refined version of an ordinary kitchen blender. Cells may be burst by osmotic shock, that is, by leaving them in a hypotonic solution so that they absorb water. Cells may also be burst by putting them under great pressure in a hydraulic press, then suddenly releasing the pressure. The Mickle Shaker agitates a mixture of cells and glass beads at up to 3000 oscillations per minute, and this is so violent that even the organelles may be broken up. Ultrasonic sound can also be used to shatter cells. Sometimes two different methods can be combined in one machine, such as the Kontes Disintegrinder.

Once homogenised, the cytoplasmic soup is separated into convenient fractions by a series of centrifugings, called **fractionation**. Generally, the material is first suspended in a solution such as 0.25 M sucrose. Early centrifugings may last only 10 min at 1000*g*, but to isolate small particles such as microsomes 100 000*g* may be needed for 30 min. After each centrifuging a fraction is removed as a pellet and is washed several times as already described (see p. 64). To check the final purity of a fraction, it can be examined under the electron microscope. Alternatively, it can be tested biochemically; for instance, its enzyme activity can be measured.

There is a serious drawback in many experiments; sophisticated preparation of cells or tissues may mean that they no longer bear much resemblance to a living organism (see p. 28). This problem is particularly acute in homogenising and fractionating. Quite apart from the obvious violence involved, most of the methods generate heat, so that the temperature of the soup may rise considerably. No one needs reminding about the effect this can have on enzymes. The pressure, too, can be pushed up to enormous values, not only in homogenising machines but also in centrifuges; a centrifugal force of 1000*g* causes a pressure rise of about 10^5 Pascals (Pa) or one atmosphere, so the final stages of fractionation may produce pressures of 100 atmospheres. It might be worth considering what effects this could have on organelles. What would happen to you at 100 atmospheres?

This whole question of damage to material raises a serious doubt about measurements on isolated mitochondria and other fractions. How do we know that the mitochondria, after going through the ordeal of being isolated and purified, behave or even look as they did in the living organism? Often there is no easy way of telling. Sometimes all one can do is to accept measurements which look reasonable, and reject others that could be artefacts due to damage. Hillman's book on biochemical techniques explores these questions in detail. They also bedevil the use of the electron microscope, which is the subject of chapter 8.

7 Seeing with light

I hope some readers of this book are naturalists. If so they will know what enjoyment can be had from watching living things in their natural surroundings. No tools are needed, just time and a pair of eyes. Watching a spider building its web, or a liverwort capsule bursting to release its spores, can be fascinating and also beautiful.

But the naturalist who likes smaller organisms can see a lot more with a hand lens. I have one magnifying × 20; it is a little difficult to use, but it means that I can see the cells in a moss leaf, and can gaze into a jumping spider's staring eyes on equal terms.

This and the next chapter are all about seeing very small things. At the outset it is important to make a distinction between magnifying and resolving. **Magnifying** simply means making bigger. Suppose a photograph were taken of an elephant 50 metres (m) away. On the negative, the elephant's image might be only 1 cm high, but if we had a large enough piece of paper we could print from it a life-size picture of the elephant. Would the photograph show every hair and flea that was on the elephant? Very likely not. In fact, it would probably be impossible to see any more detail on the life-size picture than could be made out on a print measuring only 20 cm. All that would happen is that the graininess of the negative would become more depressingly obvious as larger prints were made.

Resolution

Suppose you look at a moss leaf through a hand lens. The main reason why a hand lens is useful with a moss leaf is not really that the leaf looks bigger, but because more detail can be seen. The naked eye cannot detect the cells because they are too close together; another way of saying this is that the eye cannot **resolve** them. The hand lens helps us to resolve them. In other words, it enables us to see a separate image from each cell.

Under normal conditions our eyes can separate details which are only 0.1 millimetre (mm) apart, or perhaps a little closer. This is the limit of the eyes' resolving power. However, microscopic distances are very much smaller, and so different units of measurement are needed.

These are shown in figure 36, together with representative objects with which you will be familiar. One thousandth of a millimetre is a micrometre, or μm. This is convenient for measuring cells, mitochondria,

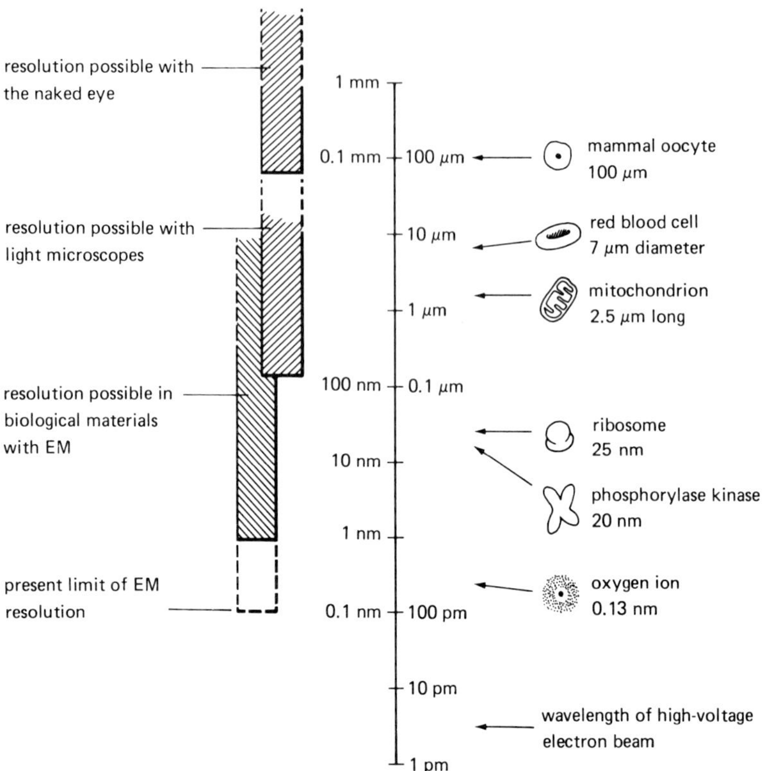

Figure 36 Logarithmic scale of sizes from 1 mm down to 1 pm. Representative objects are shown on the right-hand side with typical measurements. EM: electron microscope.

bacteria, and viruses. Going even smaller, a thousandth of a micrometre is the nanometre, or nm; this is used for things ranging in size from ribosomes down to molecules. The Ångstrom unit is 0.1 nm.

Biologists generally find these units are adequate for their purposes. However, when atoms and electrons need to be described, an even smaller unit is used. This is the picometre, or pm. It is one thousandth of a nanometre.

Lenses

When we are seeing things with light, the instrument which improves resolution is the **lens**. This works by collecting light coming from an object, and converting it into a pattern or image which looks just like

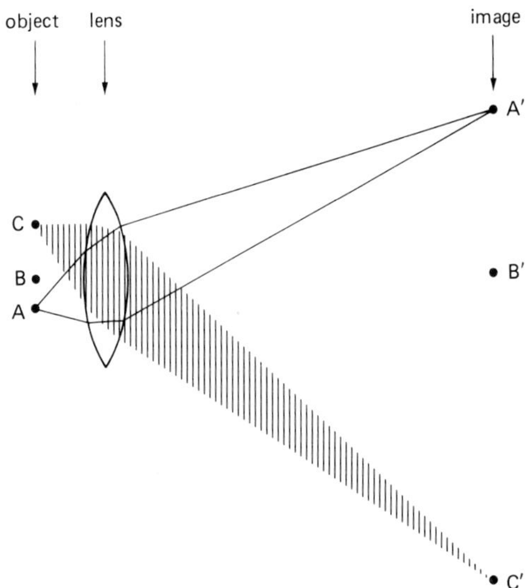

Figure 37 The formation of an image by a glass lens, seen edge-on. A, B, and C are three points on the object being examined. Hatching shows how light from C is concentrated at point C′ on the image. The same for point A is shown in black outline.

the object. The effect is produced by bending the light rays as they pass through the lens. In figure 37, three points on the object each give off light in a variety of directions. All the light from point A that hits the lens is bent in such a way that it comes to another point at A′. The same happens to the light from B and C. If a screen is placed along the line A′B′C′, a magnified image of the object is seen. The three points on the object might be three cells in a moss leaf, so close together that the naked eye cannot resolve them; but if they are more than 0.1 mm apart in the image, they will now be resolved.

One might think that just by perfecting the lens system, one could go on improving resolution indefinitely. Unfortunately this is not possible. If point A on the object is a very small spot of light, it is found that at high magnifications A′ does not look the same. The image of a spot of light has a distinctive pattern called **Airy's disc** (figure 38a). It consists of a bright patch in the centre, surrounded by alternating light and dark rings. These are caused by diffraction, which is an inevitable consequence of the wave nature of light. So the rings cannot be eliminated, and the central patch cannot be made smaller than a certain size.

The radius *R* of the central patch is fixed by the wavelength of the light used. Because the wavelength of visible light cannot be shorter than

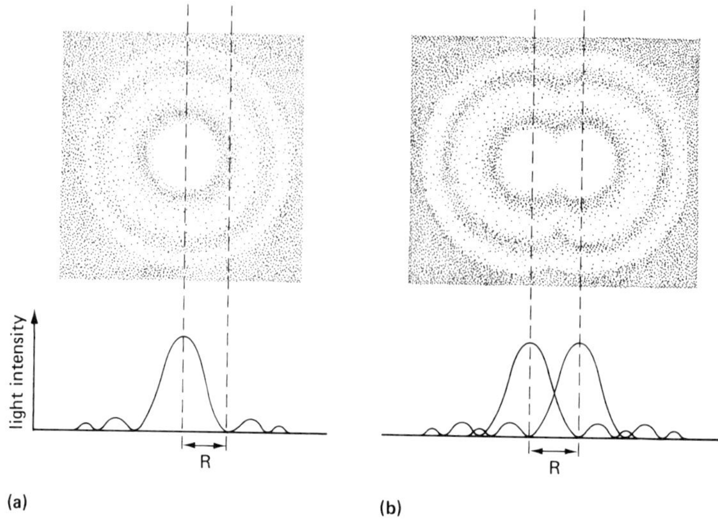

Figure 38 Resolution. **(a)** Airy's disc, the highly magnified image of a spot of light. Below it is a graph of the light intensity along a line drawn across the pattern. **(b)** Two Airy's discs distance R apart, at the limit of resolution.

about 0.4 μm, the radius R cannot be smaller than 0.2 μm. If two points on the object were so close that the centres of the Airy's discs were distance R apart, they would only just be distinguishable (figure 38b), so the value of R determines the resolving power. In other words, the light microscope cannot resolve any two points closer than 0.2 μm. Incidentally, the word 'resolution' is being used here in just the same way as in the context of chromatography (see p. 52); two patches on a chromatogram are said to be resolved only if they can be distinguished from each other. As in the chromatogram, the resolution of a microscope or lens is a measure of its sensitivity.

The light microscope

Despite the spectacular results achieved with electron microscopes, the light microscope has by no means been abandoned. Not the least of its advantages is its comparative cheapness; we are not yet in the age when every school or even research laboratory has its own electron microscope. Another important advantage is that the light microscope can be used to observe living organisms and cells. With the electron microscope this is very difficult.

A microscope made simply of the three lenses shown in figure 39 would not work very well. Single glass lenses suffer from two major problems.

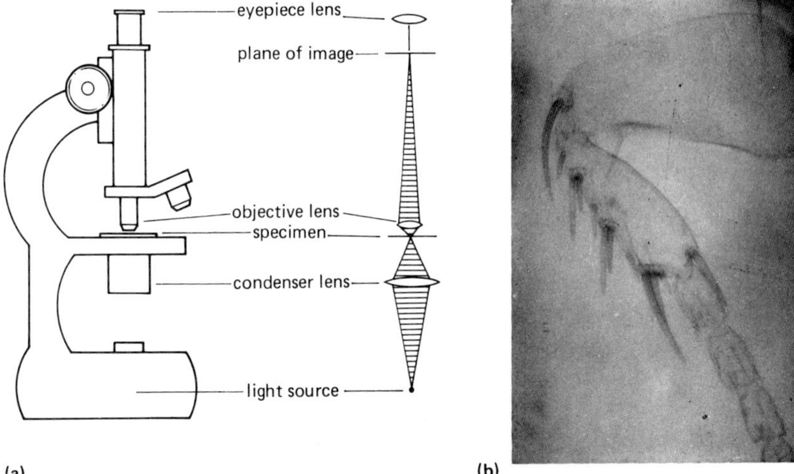

Figure 39 (a) A light microscope with the simplest possible arrangement of lenses. Light from a lamp or mirror is focused by the condenser on to the specimen. Light from each point on the specimen is brought to a focus in the image plane by the objective. The image is viewed and further magnified by the eyepiece. **(b)** A flea's leg seen by ordinary transmitted light.

In the first place, light passing through the edge of the lens does not focus at the same spot as light going through the centre. This causes a distortion called **spherical aberration**. Secondly, light rays of different colours focus at different points, producing **chromatic aberration**. To cure these defects, the condenser, objective and eyepiece are each constructed from a number of lenses made of different kinds of glass; so each is in fact quite complex, though it will be shown as only a single lens for the sake of clarity.

Condensers fulfil two functions. Firstly, they focus light from the sun or a light bulb on to the object. To check that the condenser is set up correctly to do this, it should be possible to see the surface of the light bulb, or the ground-glass screen if there is one, by looking down the microscope. A pin held against the bulb or screen enables the focusing to be checked if there is no printed matter on it. In practice, the condenser is often kept slightly out of focus to avoid having '20 watts' or 'Made in Britain' stamped across every specimen being examined.

The second function is to illuminate the specimen from below, evenly and through quite a wide angle. This angle should be fixed so that the objective lens, but nothing outside it, is filled with light (figure 40a). To achieve this, the condenser diaphragm is gradually closed until the field of view just begins to get darker. It is best to do this with no specimen

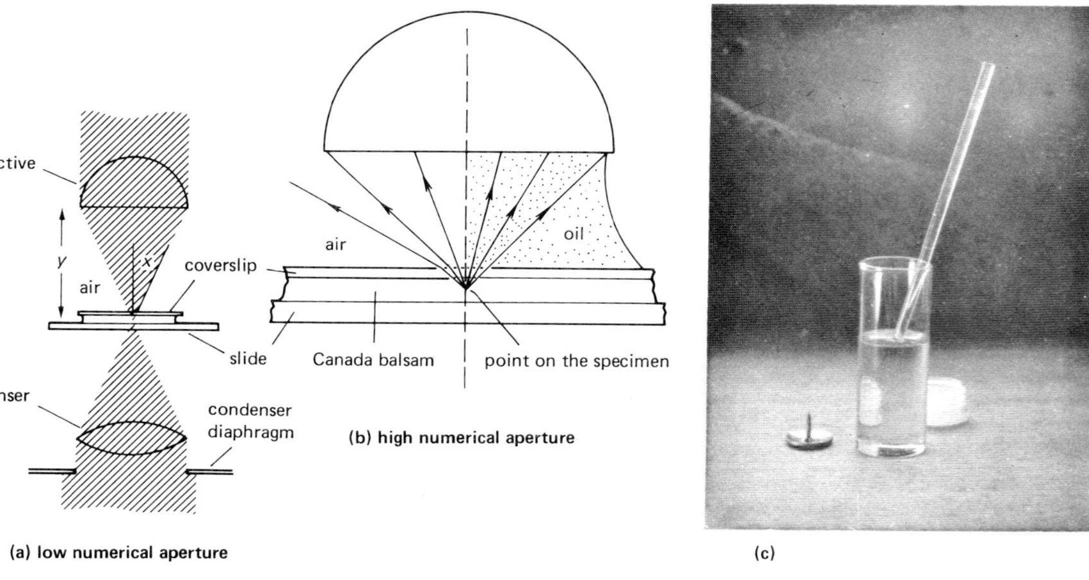

Figure 40 How light reaches the objective from the specimen. **(a)** Light passes through a medium of low refractive index (air) and a small angle *x*. *y* is the working distance. **(b)** A high-power objective, showing use in air on the left, and oil immersion on the right. Three light rays are drawn on each side. **(c)** A glass rod standing in a jar of immersion oil.

on the stage, so that light from the condenser passes straight through the focus point to the objective lens. Rays taking this path, and not affected by the specimen, are known as **direct rays**. Direct rays are important, since some small objects are visible only because the direct rays interact with other rays scattered by the object.

Any resolution achieved by the microscope depends on the **objective** lens. Magnification at the objective is × 100 or so, at the most, and it produces an image a short distance below the eyepiece (figure 39). The light passing on up the tube from the objective contains all the detail that can eventually be seen by the eye. So, the quality of the microscope as a whole can clearly be no better than the objective. If our eyes were good enough, we could examine the image produced by the objective and immediately see all the detail that the objective had managed to resolve in the specimen.

However, our eyes are not that good. The **eyepiece** is there in order to magnify the image made by the objective. This means that we can then see all the detail the objective has resolved. Once a strong enough eyepiece has been found to do that, there is no point in using even stronger ones. Although they will certainly make the image larger, they cannot improve the resolution.

Numerical aperture

The size of Airy's disc, and therefore the resolving power, depend not only on the wavelength of the light. Lens design is important, and in particular the way light enters the objective lens from the specimen. **Numerical aperture** is the aspect of the lens used to describe this. Other things being equal, a higher numerical aperture means greater resolving power. Numerical aperture is defined as $r \sin x$, where r is the refractive index of the medium just below the objective lens, and x is the angle between the vertical and the outermost ray entering the lens (figure 40a).

A material's **refractive index** is a measure of the speed of light passing through it. Light travels through air at about the same speed as through a vacuum; the refractive index is 1. Even though other substances are perfectly transparent, they may have an important effect on light which we cannot always detect with the naked eye. Light is considerably slowed down passing through glass, and to express this we say that glass has a high refractive index, about 1.5. We cannot see this effect when the light passes through the glass at right angles to the surface. However, it can be detected when the light approaches the glass at some other angle. In fact, a change in refractive index causes the light to bend. This is how glass lenses in air can be made to bring the light, coming from a point, to focus at another point on the other side of the lens (figure 37). The ability of some transparent materials, such as protoplasm, to slow down the light passing through them is the foundation of the phase-contrast method, which is described later in the chapter.

Explaining why the angle x is important for resolution would involve a detailed excursion into optical theory. Briefly, the light from any point on the specimen spreads out in all directions. Having a short working distance y and a large angle x means that more of this light can be collected by the objective lens. One advantage is, of course, that there is a greater quantity of light from the point, so that a brighter image of it should result. But in addition, the light rays leaving the point in different directions contain different kinds of information about the point. Therefore, as many of these rays as possible must be collected; otherwise the image is distorted and the microscope is not accurate. The fine details of the image depend on information contained only in the light which leaves the specimen at high values of the angle x.

Oil immersion

Oil immersion is a technique designed to increase the numerical aperture to as high a value as possible. One reason why the oil helps is that its refractive index is much higher than that of air. In fact, it is specially mixed to have a refractive index the same as glass, 1.5. Therefore, a glass rod standing in a bottle of immersion oil will be easily visible above the oil surface, but should be invisible below it (figure 40c). The Canada balsam in which specimens are often mounted also has about the same refractive index. So, as seen in figure 40b, light rays travel in straight lines from the specimen to the objective if the substances in between are Canada balsam, glass and immersion oil.

There is more to the method than this, however. On the left side of figure 40b, the light rays will bend when they pass into air from the coverslip. As a result, the outer ones bend away from the objective and miss it altogether. Using oil, the same rays are not bent, but carry on in a straight line to enter the objective. It can be seen that, in oil, light is collected from a point on the specimen through a wider angle x than in air. Therefore, more information is collected from the point, as well as more light; hence a more accurate image and better resolution are possible.

Oil is needed for the highest-power objectives. Focusing should be done very carefully, since the working distance is short and the oil makes it difficult to see how far the objective is from the coverslip. Afterwards the oil should be carefully wiped away using clean, dust-free, lens tissue. This is important because dust acts as an abrasive; it can easily cut little grooves in expensive lenses.

Preparing the specimen

Many of the advances in microscopy have resulted from new designs of lenses and optical arrangements, but almost as important have been innovations in the preparation of specimens. In the seventeenth century Robert Hooke published his famous book *Micrographia*, and for these microscope studies the lighting always came from the side or above the specimen. Only reflected light passed into the lenses. Then in the nineteenth century it was a popular pastime to watch pond life using **transmitted light**. In this arrangement the object being studied is between the light source and the objective, as in figures 39 and 40. However, these two methods by themselves allow us to examine only a limited range of objects.

Maceration is a simple technique which yields useful results. Tissue is treated in such a way that the individual cells separate from each other. Then the shape and internal structure can be seen without recourse to sectioning.

However, **sectioning** has for long been the main way of 'getting inside' tissues. Thin sections of a uniform thickness were not possible until the invention of the rocking microtome in the last century. It is not possible to shave very thin slices off a piece of rubbery or fragile tissue just by using a sharp knife. It would be like trying to cut a thin layer away from a bath sponge with a razor blade. So, to make a section for microscopy, the tissue needs to be embedded in a block of wax. Then a sharp knife can easily take a thin layer off the surface of the block, while the tissue is firmly supported inside the wax.

When a bacon slicer carves a joint of bacon, the lump of meat is carried sideways past the blade. Then it moves back to its original position, is pushed forwards a short distance, and once again moves sideways past the blade. If it is moved forwards by 2 mm, the slice cut off is 2 mm thick. A **rocking microtome** works in the same way but on a smaller scale. For light microscopy, sections 5–10 μm thick are needed, so this is the distance the wax block is moved forwards each time a section is cut off.

A great deal of information can be gained by **staining** the section. The most useful stains are specific for certain constituents of the protoplasm. For instance, toluidine blue stains only RNA, and Feulgen stain is specific for DNA. Different-coloured stains specific for different materials can be used in combination. For example, one method using four stains makes cartilage bright red, nuclei purple, connective tissue green, muscle light red, submucosa orange, and so on. This is not only pretty but informative too.

Most stains are toxic to living cells, or will only work satisfactorily on dead material. This could be a serious drawback, so **vital stains** are particularly useful. These are stains which are taken up by living cells without apparent damage to them. An example is methylene blue. However, most satisfactory of all are the biological materials which are coloured in their natural state. There is of course no need to stain chloroplasts to see where they are.

How are things seen?

There are several ways of seeing with light microscopes. Reflected light can be collected from the specimen if it is lit from the side or above; and if transmitted light is used, absorption by one part of the specimen will make it appear darker than other parts absorbing less. The use of stains makes some wavelengths of light pass through the specimen more strongly than other wavelengths, and this gives rise to colours.

There is in fact a considerable range of other approaches to making images, in both light and electron microscopes. In the rest of this chapter we shall look at two more methods using light. They use, respectively, dark ground and phase contrast. Both are ways of making observations on living cells which cannot be made easily by another means. Usually this is because the objects to be observed are too transparent to be seen in a living state in any other way.

Dark ground

Dark-ground illumination is well named. If you look down the microscope before putting the specimen on the stage, nothing at all should be visible. Then when the specimen comes into view it appears as a silvery object on the black background.

The reason for the dark ground is that all the direct rays (see p. 79) are cut out (figure 41). Only light reflected or scattered from the surface of the specimen contributes to the image. So in this way it is something like using top or side lighting, though in fact the light comes up through the slide as in ordinary transmission microscopy. In its simplest form, the condenser is modified by blocking out the central zone of light. An opaque disc is all that is needed. It must be positioned so that a hollow cone of light focuses on the specimen, but then continues past the objective lens, not entering it. In this way the object, often a diatom or a protozoon, is illuminated evenly all round but obliquely from beneath. No matter how transparent it is, it should show up if its surface is capable of reflecting or scattering light.

In some ways the opaque disc is the exact opposite of the condenser diaphragm. The latter stops all the direct rays except those entering

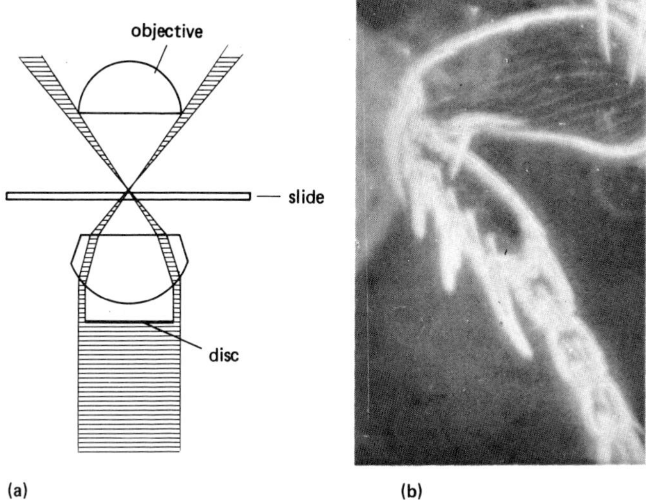

Figure 41 Dark-ground illumination. **(a)** A simple illumination system. **(b)** A flea's leg seen with the dark-ground method; it is the same specimen as seen in figure 39.

the objective, while in dark-ground illumination the disc stops all those that would have reached the objective. For high-power objectives, a more complicated condenser is used, but it achieves the same purpose.

Phase contrast

Unfortunately dark ground is no use for displaying chromosomes; they do not reflect light adequately from their surfaces. However, excellent results can be achieved with the **phase-contrast** method. This was only developed in the 1940s, and it represents a much more revolutionary way of seeing transparent objects. It makes use of the fact that cells, organelles, and the medium they are in may have different refractive indices.

Light is often said to be a wavelike motion, or to travel in waves. Figure 42a shows such a wave. However, this shape does not mean that something actually moves up and down, like the surface of a choppy sea. The wave form is best thought of as a graph with time along the horizontal axis. According to the electromagnetic theory of light, light consists of disturbances in the electric field and also the magnetic field at any point the light passes through. This is a hard concept to grasp, and fortunately for present purposes there is no need to. The same graph can equally well represent the state of the electric (or magnetic) field along a line at a given instant. Taking both graphs together gives quite a good idea of what is happening along the path of a light ray.

Amplitude is the vertical distance from peak to trough. It is important to us because greater amplitude means greater brightness. Suppose now that two light waves arrive at the same point in space. If both waves' peaks and troughs coincide, the disturbances will reinforce each other (figure 42b). It will be the same as if a wave with much greater amplitude had arrived. In other words, the light will be brighter than if only one wave was present.

If, however, the waves are separated by half a wavelength, they are said to be exactly **out of phase** (figure 42d). Now the result is quite different. The peak of one wave is at the same place as the other wave's trough;

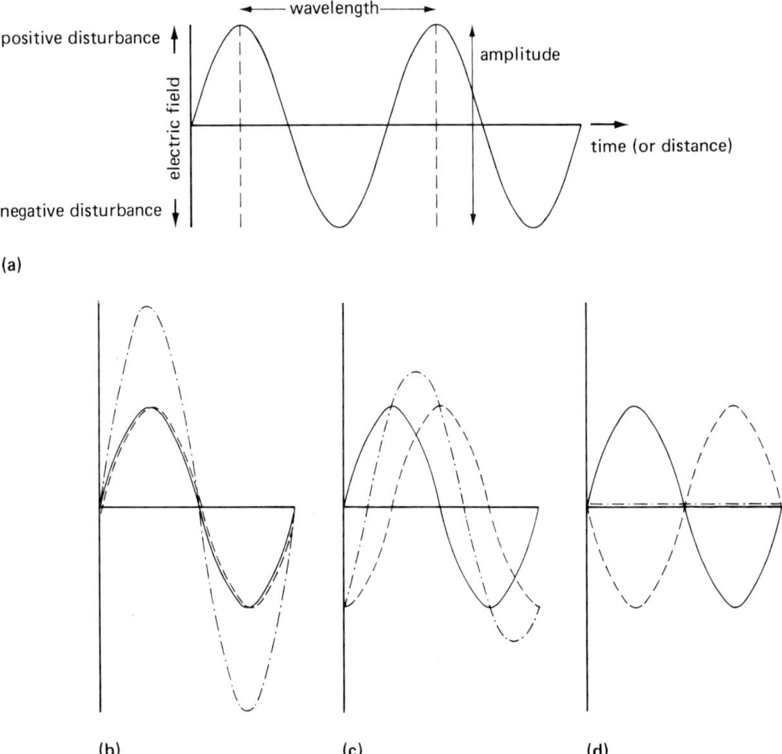

Figure 42 **(a)** The wave form of light, interpreted as a graph of the changes of the electric field at a point the light is passing through. Alternatively, the same shape can be interpreted as a graph of change with distance at an instant of time. **(b–d)** Interference between two waves shown as continuous and interrupted lines. The result of adding the two waves is the dot-and-dash line. The two waves are **(a)** in phase, **(b)** out of phase by a quarter wavelength, and **(c)** out of phase by a half wavelength.

so if they have the same amplitude they cancel each other out. In fact, the waves cancel each other out all the way along. Now the amplitude has become zero, and so no light is seen at all. This effect of the waves on each other is called **interference**. In the intermediate situation, with the waves separated by only a quarter of a wavelength (figure 42c), the light has not disappeared. In fact, it is actually brighter than one wave on its own, but not as bright as when the waves are exactly in phase.

This is an important result, because the situation in figure 42c often occurs when light passes through a small specimen. It was said earlier that a high refractive index means that light is slowed down passing through the substance. So, when a *Paramecium* or an *Amoeba* is examined in water under a microscope, the direct rays go past the cell in the water, while the other rays go through the protoplasm (figure 43a). If the refractive index of the protoplasm is higher than water, the second lot of rays will be retarded a little. Often this retardation is about a quarter of a wavelength.

Sometimes a small specimen is visible under the microscope only because of the interference of the two sets of rays. When they come together in the image, any interference will mean the brightness diminishes. But, as we have seen, a retardation of only a quarter of a wavelength does not have a great effect. It may not be enough to make transparent objects like chromosomes visible at all.

The whole point of the phase-contrast method is to retard the indirect rays until they are about half a wavelength behind the direct rays. Full interference should then be possible, making parts of the transparent specimen noticeably darker. The trick is to pass the indirect rays through some other substance so that they are retarded again. To do this it is necessary to have a source of light shaped like a ring; a device similar to that in figure 41 is used, except that in this case the direct rays enter the objective. The condenser produces an image of the ring shape inside the microscope.

Exactly at the point where the image is formed, a specially shaped glass disc is fixed, called the **phase plate**. It is the same thickness all over except for a groove which is cut in a ring shape (figure 43b). The image of the light source is made to fall on the groove, and this means that all the direct rays pass through the thin glass. On the other hand, most of the indirect rays are scattered by the specimen, so that they pass through the rest of the phase plate. This means that they have to cross thicker glass, which retards the indirect rays more than the direct rays. The depth of the groove is calculated to make the extra retardation about a quarter wavelength.

Not only does the phase plate change the phase relation of the two sets of rays; but the full interference effect only comes about if the bright-

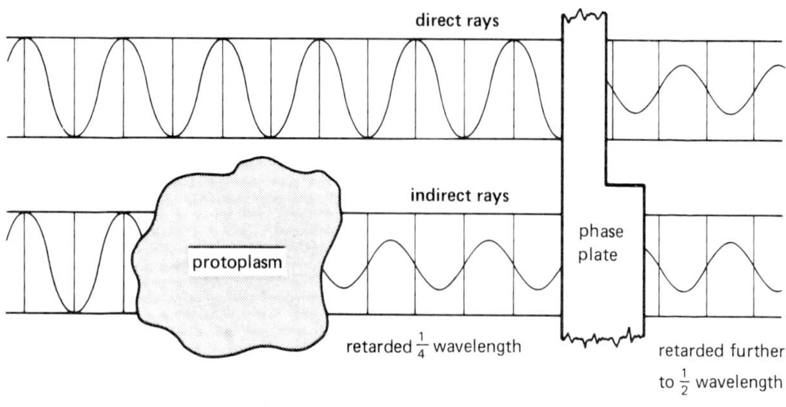

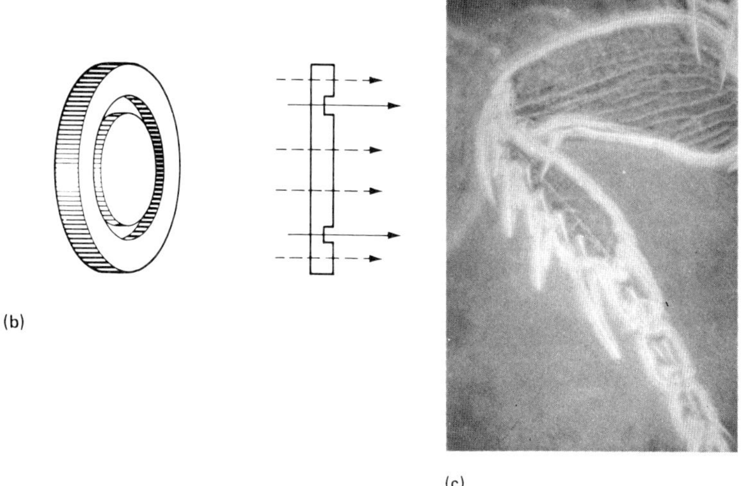

Figure 43 Phase contrast. **(a)** Rays passing through the specimen are scattered and retarded by about a quarter wavelength. At the phase plate they are retarded by another quarter wavelength relative to the direct rays. **(b)** Phase plate, from the side and in section; the direct rays (continuous lines) pass through the groove while the indirect rays (interrupted lines) from the specimen are retarded by passing through the thicker glass. **(c)** A flea's leg seen by phase contrast; it is the same specimen as in figure 39.

ness of the direct and indirect rays is about the same. Scattered indirect rays are usually much dimmer than the direct rays, so the phase plate is used to reduce the brightness of the direct rays. This is done simply by coating the surface of the groove with a material which absorbs some of the light striking it. Ideally, the result should be as shown on the right side of figure 43a. When the two sets of rays recombine in the image, parts of the transparent specimen which are normally invisible should appear dark. By this means, impressive time-lapse films of mitosis have been made, something which is impossible using a stain, since this kills the cells.

8 Seeing with electrons

Biologists today are quite rightly excited by the revelations of the electron microscope. It is difficult to imagine what biology was like without it. Indeed, one could seriously argue that the modern era of biology dates from the early 1950s, with the DNA double-helix theory appearing in 1953 and the invention three years earlier of the glass knife (figure 45).

Light and transmission electron microscopes are similar in general principle. If the light source is replaced by an electron gun, the glass lenses by electromagnetic lenses, and if a fluorescent screen is used in place of the naked eye, we have an electron microscope (figure 44).

The electron gun

In a normal microscope lamp, the light is emitted from a filament heated by an electric current. The source of the electron beam, or the **electron gun**, is also a heated filament. It is a bent piece of tungsten wire. One advantage of tungsten is that it can be heated to 3000 °C without melting, and at this temperature plenty of electrons fly off its surface. The importance of the bend is that the temperature, and hence the electron emission, is most intense at that point.

Just as in the mass spectrometer, the electrons must be accelerated to form a beam. This is achieved by positioning a positively charged anode, with a hole in it, a short distance from the tungsten filament or cathode. A charge difference between the cathode and anode, usually of about 40 kV, accelerates the electrons to an enormous speed. A kilovolt (kV) is 1000 volts. The 40 kV difference is the **accelerating voltage.** Voltages as high as 100 kV can be used in ordinary electron microscopes, and one advantage of this is that a shorter wavelength is produced.

Powerful though such an electron beam is, it would not get far in ordinary air. Very soon, nearly all the electrons would collide with gas molecules. If most of the electrons are to reach the far end of the microscope, there must be a vacuum inside the instrument. This is unfortunate, because it means that living specimens cannot be examined in most electron microscopes; they would soon become desiccated because their water would evaporate off. Certain high-voltage instruments can be used with living material, and these are described later in the chapter (see p. 99).

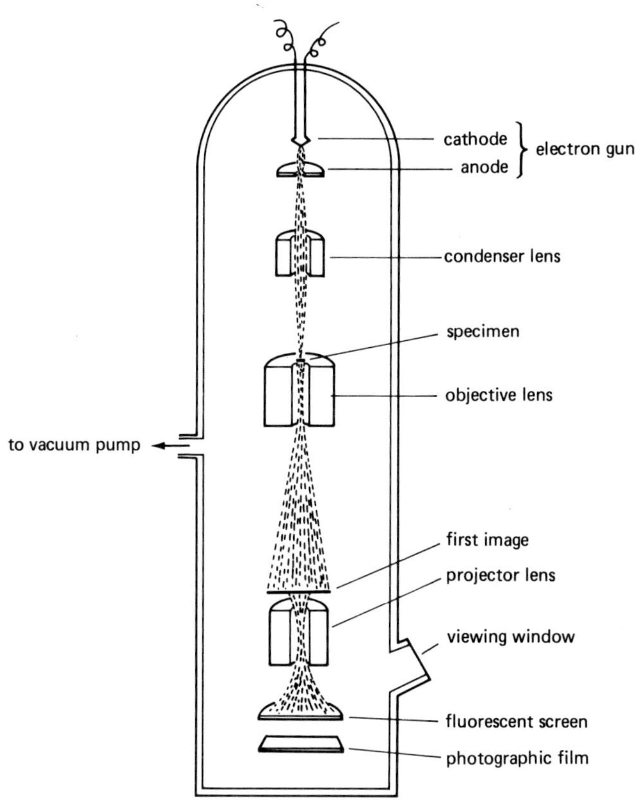

Figure 44 Transmission electron microscope. The lenses and other components are shown cut in half.

Focusing electrons

The great advantage of using electrons rather than light is that much better resolution is possible. Because Airy's disc cannot be reduced below a certain size (see p. 75), an ordinary light microscope can only separate details more than 0.2 μm apart. Using ultraviolet light of short wavelength, we can improve resolution to about 0.1 μm, but that is the limit using light.

It may seem strange to think of electrons as radiation like light, since for many purposes it makes sense to regard them as negatively charged particles. But an electron beam behaves in many ways like a light beam. Most important, it has a wavelength which is extremely short. If electrons are accelerated by a voltage difference of 100 kV, the wavelength is only

4 pm (figure 36). That is one hundred thousandth the wavelength of light. This means that Airy's disc can be made much smaller than with light.

A glass lens would be useless for bending a beam of electrons because, apart from anything else, glass is opaque to them. Instead, **electromagnetic lenses** must be used. Just as a beam of ions in a mass spectrometer (see p. 37) can be bent by a magnetic field, so also can a beam of electrons. In particular, it can be brought to a focus at a point, just like a beam of light passing through a glass lens. The magnetic field is created by a small electric current flowing through a coil of wire with many turns.

If electromagnetic lenses had the same kind of numerical aperture as glass lenses, the resolution achieved by electron microscopes could be fantastically good: around 2 pm, perhaps. Unfortunately, the lenses are nothing like as efficient as glass lenses, and the best resolution achieved so far is about 0.2 nm. This is sufficient for most purposes, and has enabled the atoms in a crystal of gold to be clearly resolved. However, biological materials are more difficult to handle, and in practice the best resolution obtained is 2 to 3 nm.

Electromagnetic lenses are obviously very different from glass lenses, and one further difference is of some use. In a light microscope, changing the magnification usually means changing the objective or eyepiece lens. But there is no need for this in the electron microscope; this is because the magnification of the lens can be increased simply by increasing the current flowing through the coil.

The transmission electron microscope

Figure 44 shows a simplified layout of an ordinary electron microscope. In many ways it is just like a light microscope. The electron beam is focused onto the specimen by one or more condenser lenses. This gives a bright illumination; although it is in fact electron bombardment of the specimen, it is still called 'illumination'. Electrons leaving the specimen then continue on down the column of the microscope, passing through one or more objective lenses to give the first image. It is the quality of this image, as achieved by the efficiency of the objective lens, that determines the microscope's resolution.

However, that image is not used. The electron beams enter yet another lens, called the projector, which focuses the beam again at a second image. Electrons are invisible to the naked eye, and it would be dangerous to look at them anyway. So, for routine examination they are made to strike a **fluorescent screen**. This is a sheet covered with zinc sulphide or some other material which will fluoresce; that is, when electrons strike it, light is given out. The screen converts the pattern of focused electrons into a visible image. Fluorescent screens do not allow particularly good resolution, so anything worth detailed examination is photographed. To do this, the screen is moved out of the way and the image then falls onto a film underneath.

Forming an image

Just as in the ordinary light microscope, some of the illumination striking the specimen passes right through it. This is what is meant by '**transmission**'. For any electrons at all to be transmitted, the specimen must be very thin indeed. If it is a tissue section, it must be at most 100 nm thick for ordinary instruments. But the trouble with an untreated section of tissue is that if electrons can get through it in one place they can get through it everywhere. This is because most of the atoms in the tissue are carbon, hydrogen, oxygen and nitrogen. These are all fairly transparent to high-speed electrons.

However, heavy-metal atoms are very different; they scatter electrons, making them fly off in various directions. To get a useful image of a thin section, the aim is to add heavy-metal atoms in such a way that they show up the structure in the specimen. Suppose, for instance, that they are absorbed by membranes and not by the rest of the cytoplasm. Only the electrons which are not scattered enter the objective lens and so contribute to the final image. The membranes therefore appear as dark lines on the final image. The heavy-metal atoms are functioning just like the stains in light microscopy.

Preparing specimens

Specimens must go through a complicated preparation before they are placed in the microscope.

The first thing to be done is to **fix** the tissue. As soon as cells die they start to disintegrate. To stop this, tiny pieces of tissue about 1 mm^3 are cut out and dropped into a fixative solution. In electron microscopy, the usual fixatives are aldehydes, osmium tetroxide, and occasionally potassium permanganate. Fixatives seem to work by linking together molecules in the cytoplasm with chemical bonds to form an insoluble network. However, only some parts of the cytoplasm are affected. Often glutaraldehyde is used, followed by osmium tetroxide, but even then only proteins and lipids are fixed. This may amount to only 20 or 30 per cent of the tissue; all the rest will be washed out or destroyed later in the preparation.

The tissue must next have all its water removed; in other words it must be **dehydrated**. This is not only because the microscope needs a vacuum, but because to make sections the tissue must be embedded in plastic.

Plastic is used instead of paraffin wax (see p. 81) because wax does not allow really thin sections to be cut. If the tissue had not been dehydrated the plastic would not soak in, since it is not soluble in water.

Just as in light microscopy, dehydration is usually done by passing the tissue through progressively stronger solutions of ethanol: 70 per cent, 90 per cent and finally 100 per cent. The trouble with ethanol is that it dissolves out lipids from the tissue as well as removing water, so dehydration must be done as quickly as possible. Unfortunately, the plastics used for embedding also remove lipids.

Sectioning and staining

Metal knives are not good enough for this kind of sectioning and diamond knives are sometimes used. However, most electron microscopists make their own knives from small pieces of plate glass (figure 45). These are carefully broken along a diagonal to give a top edge 5 or 6 mm long. Although the edge on a good knife will be extremely sharp, it must be used within a day or two. This is because glass is supercooled liquid; within a short time it simply flows out of shape.

Sections for electron microscopy need to be between 10 and 120 nm thick. For most purposes 50–60 nm is best. The sections are extremely delicate, so they cannot be picked up with forceps. Instead, a trough to hold a drop of water is fixed to the sloping surface of the knife. The specimen in its block of plastic is mounted in the microtome, and to cut a section the block is brought down past the knife edge. As it passes, the edge shaves off a section which floats onto the water. A useful instrument for pushing it around at this stage is an eyelash fastened to a stick.

Figure 45 Sectioning. On the left is a glass knife with the edge at the top. The knife in the middle has a trough attached, and on the right the trough has been filled with water. Circular grids for supporting the sections are seen near the drawing pin.

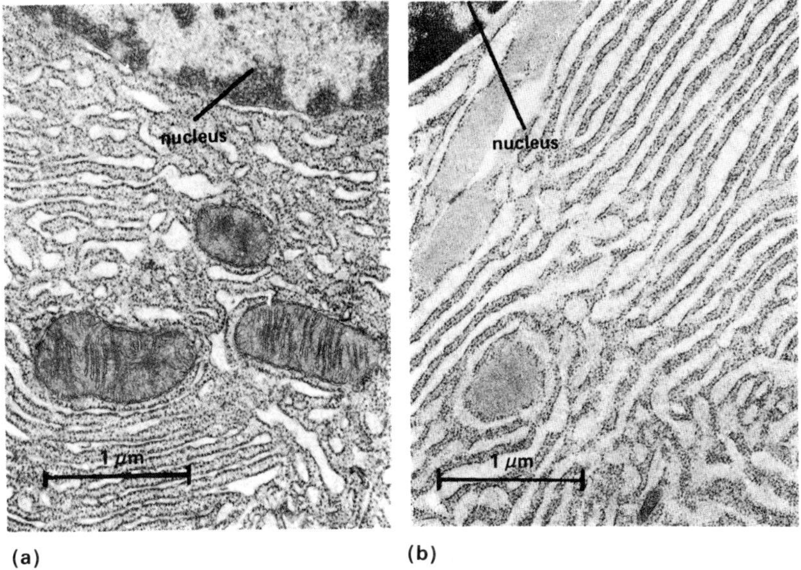

Figure 46 Transmission electronmicrographs of the same type of cell – a pancreatic acinar cell. Both sections are stained with uranyl acetate and lead citrate, but **(a)** is fixed with osmium tetroxide and **(b)** with glutaraldehyde. What differences are there between the two sections?
(**(a)** and **(b)** courtesy of B. S. Weakley; **(b)** from her *Beginner's Handbook in Biological Transmission Electron Microscopy*.)

From now on the delicate sections must be given some kind of flat support suitable for use in the microscope. A glass slide is no use, since it does not let electrons through. It would be like using a metal slide in a light microscope. Instead, a **grid** is used; this is a small disc of mesh, made of metal with nothing at all in the holes. The sections are just strong enough to avoid collapsing into the holes. The metal will scatter or absorb electrons in the microscope, but the bits of section showing through the holes will transmit the beam to give an image.

The sections are carefully transferred to a grid in a drop of water. One way to do this is to immerse the grid in the water and bring it up under the sections. The water is then removed by blotting. Once attached to the grid, the sections are stained by dipping the whole grid into a suitable solution. To stain only one surface of the section, the grid can be floated on the stain, section side down. This makes the section appear, in the electron microscope, even thinner than it really is.

Two common stains are uranyl acetate and lead citrate. In these the heavy metals are uranium and lead respectively. Both stains have been used in the cells shown in figure 46. Sometimes a substance can act as a stain as well as fixing the tissue; an example is osmium tetroxide. This has been used to fix and stain figure 46a, while for figure 46b glutaraldehyde was the fixative.

If you examine the two micrographs carefully, you will soon see some important differences. With glutaraldehyde the nucleus is more heavily stained than with osmium tetroxide. The spaces or cisternae in the endoplasmic reticulum and around the nucleus are expanded more widely, and the nucleus has shrunk. Cristae are difficult to see in the mitochondria. Finally, the endoplasmic reticulum tends to run in long, parallel lines.

You may like to think about whether this is a controlled experiment. Whether it is or not, it suggests that the appearance of a section may be quite seriously changed by the method of preparing the tissue. This is mentioned again at the end of the chapter (see p. 104).

So far we have been considering the standard method of using stained ultrathin sections. However, the electron microscopist can turn instead to several ingenious methods not used in light microscopy. Three of these are negative staining, shadowing and freeze etching.

Negative staining

As its name implies, **negative staining** means staining everything except the structure you want to look at. It is particularly good for virus particles, or for DNA molecules extracted from nuclei or viruses. What these have in common is that most of the useful information can be gained from the external shape of the object, not from the internal structure.

The objects are spread out on a very thin film of plastic, and then flooded with a solution of stain. As seen in figure 47a, the stain collects around the edges of the object because of the meniscus. It also fills the cracks on the surface of the objects, and even sometimes holes inside. When the solvent is evaporated, a solid layer of stain is left, varying in thickness. The 'holes' in the stain reveal the detailed size and shape of the objects. It is rather like a fossil footprint: the foot is no longer there, and the surrounding mud has hardened into rock to show the foot's shape. In negative staining, there is no need for fixing, embedding, or sectioning.

Figure 47b shows negatively stained molecules of phosphorylase kinase. Each molecule is shaped like a butterfly and its relative molecular mass is 1 300 000. The four lobes are identical, and each one is constructed from at least four polypeptides. Phosphorylase is an enzyme which catalyses the production of glucose-1-phosphate from glycogen; but it does not function unless it is itself phosphorylated. This phosphorylation is the reaction catalysed by phosphorylase kinase. In turn, phosphorylase kinase requires calcium ions in order to function. This enzyme will be mentioned again shortly.

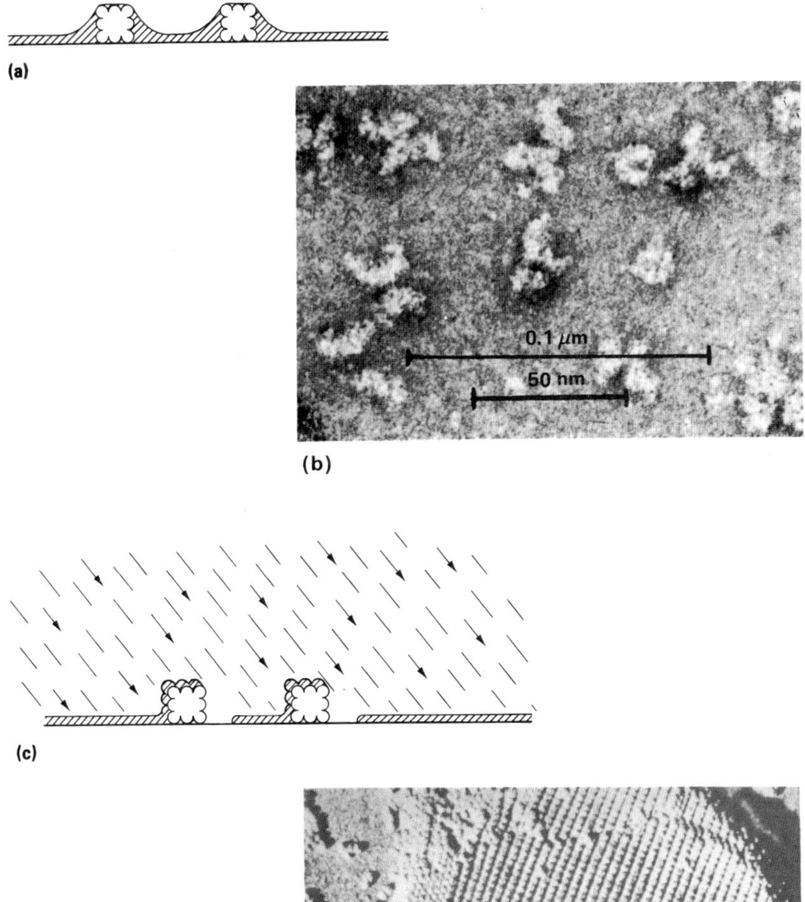

Figure 47 (a, b) Negative staining. **(a)** Side view of a preparation, showing how the stain rises up around the edges of the objects and fills the cracks. **(b)** Phosphorylase kinase molecules. **(c, d)** Shadowing. **(c)** Side view of a preparation: evaporated platinum atoms cause a layer to build up except in the objects' shadows. **(d)** Bean mosaic virus particles; some are separated but others are part of a crystal.
(**(b)** Courtesy of M. T. Davison; **(d)** courtesy of Electron Microscopy Unit, Zoology Department, Edinburgh University.)

Shadowing

Another ingenious technique is **shadowing**. This is again useful for showing the external form of objects such as viruses. The objects are scattered on a plastic film and placed in an evacuated chamber. Above them and to one side, a tungsten filament is heated by passing a current through it. The hot tip is placed in contact with a heavy metal such as platinum. At such a high temperature the platinum evaporates. Atoms shoot off in straight lines, many hitting the plastic film and the object on it. On one side of each object the atoms pile up while on the other side, in the 'shadow', there may be none at all (figure 47c).

The shadows show up on the image because they are the areas letting electrons through most easily. This means that they appear as light areas. However, when a photograph of the image is taken, a negative print rather than the usual positive print is made; this makes the 'shadows' appear black, so that they resemble the ordinary shadows cast by a light (figure 47d).

Freeze-etching

Although shadowing has proved to be a valuable method in its own right, it has attracted renewed interest because it plays a part in **freeze-etching** (figure 48). This method enables the internal structure of cells to be investigated, but without fixing, dehydrating, embedding or sectioning. No fixing is needed because the tissue is dropped straight into nitrogen at its melting point, −210 °C or 63 K. The cooling produces ice crystals in the cytoplasm, of course, but they appear to be so small as to cause no detectable damage.

Now the frozen block is split across the middle with a razor blade. The structure of the frozen cells means that the block is weaker in some directions than others, in particular along unit membranes. The block therefore preferentially splits along these lines of weakness. Next, it is placed in a vacuum so that some water sublimes, that is, evaporates from the ice crystals. This causes the surface to sink a little, leaving a few more parts of the structure upstanding. For this reason, the vacuum treatment is called 'etching'.

At this stage the block is too thick to be put into the electron microscope. In fact, the tissue is not put in the instrument at all. A replica is made of the surface by placing the block in the chamber used for shadowing. A mixture of carbon and platinum atoms shadows the surface; then a layer of carbon is deposited on it in the same way but from directly above, in order to strengthen it. Finally, the tissue block with the replica on it is allowed to warm up, and during this stage they come apart from

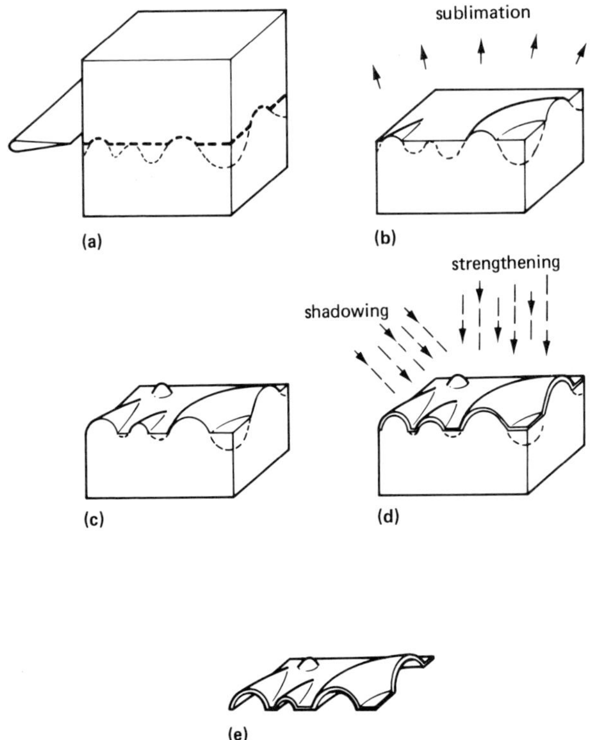

Figure 48 Freeze-etching. **(a)** The frozen block is split along a line of weakness indicated by the bold interrupted line. **(b)** The surface is lowered a little by sublimation to give the 'etched' surface **(c)**. This is shadowed and strengthened with platinum and carbon **(d)**, before the replica **(e)** is removed from the tissue.

each other. The tissue is discarded while the replica is put into the electron microscope.

Figure 49a is the result of freeze-etching a cell. A similar cell in ultrathin section is shown in figure 49b. Several features can be seen in both micrographs. Nuclear pores appear in the section as gaps in the nuclear membrane, and as round patches in the freeze-etch. It can also be seen in the latter how the outer layer of the nuclear membrane has been stripped away from three quarters of the nucleus. In the section, layers of cytoplasmic reticulum are lined up parallel to the nuclear membrane. A similar set of layers can be seen in the freeze-etch as slanting surfaces in the cytoplasm. Mitochondria also appear in both micrographs. The endoplasmic reticulum appears fuzzy in the section, away from the nucleus, probably because it has been cut obliquely.

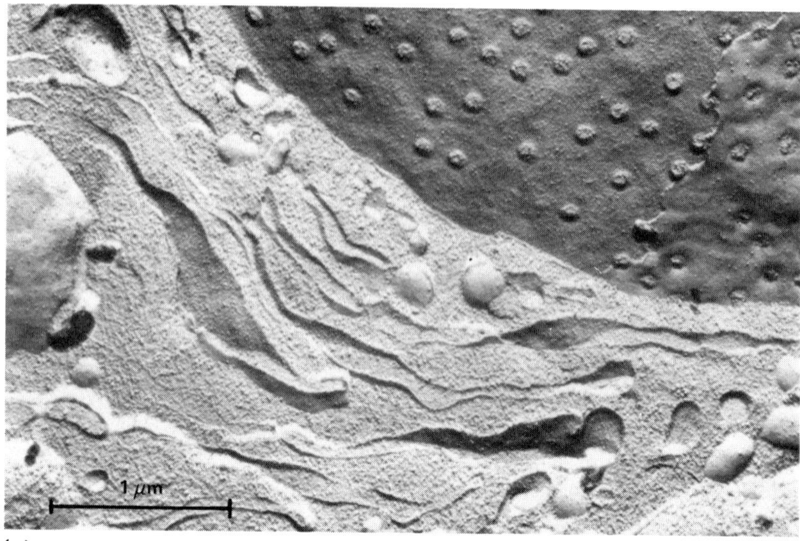

(a)

(b)

Figure 49 **(a)** is a freeze-etched liver cell, while **(b)** is a similar cell prepared as an ordinary transmission electronmicrograph. In each section, the large area in the top right corner is the nucleus. How many other structures can you recognise in both micrographs?
(Courtesy of J. B. Finean; from Michell, Finean, and Coleman (1982) *School Science Review* **63**, 434–441.)

Serial sections

Both freeze-etching and ordinary shadowing give quite a good 3-D effect. While making the micrographs interesting and dramatic, there is a more serious side to it. The problem with using stained sections is that they are so thin as to be effectively two-dimensional. A sectioned organelle may look circular, but this does not tell us what shape it is. After all, circular sections can be obtained by cutting across a tennis ball, an egg, and a plastic drainpipe. An oval section can result from other cuts through the egg and the pipe. Without further information, the shapes on a thin section do not tell us much.

One way of studying three-dimensional shapes is by using **serial sections**. This means collecting all the sections produced when the microtome slices its way through a cell or small organism. The sections are kept in the same order as they are made. A model is made of each section, or perhaps a drawing on transparent plastic. By making a pile of the models or drawings, and keeping them in the correct order, the three-dimensional shapes of the sectioned objects can be made out.

This method has led to some interesting results. For example, sections have been made of the alga *Chlorella*, well-known in photosynthesis experiments. There appear to be many small, round or oval, mitochondria on the sections. But a model made from serial sections led to a different conclusion. Unlike most mitochondria, the *Chlorella* mitochondrion is not sausage-shaped. In fact, there is only one in the cell; it is a long, branched tube, reaching most parts of the cytoplasm. It sometimes even passes through the large chloroplast in a cytoplasmic tunnel. So beware: sections of organelles are not always what they seem.

High-voltage electron microscope

A completely different approach to three dimensions makes use of our stereoscopic vision. Ordinary **stereophotography** involves taking two photographs of a scene from slightly different angles. These represent the views seen by the left and right eyes respectively. By looking at the photographs through a special viewer, they can be combined into a single image to give a convincing impression of distance.

It would be marvellous if this could be done with electronmicrographs: and indeed now it can. Various things make it possible. First, some instruments enable the operator to tilt the specimen in the microscope to different angles; this allows a left-eye view and a right-eye view to be photographed. But it is only worth doing this on sections sufficiently thick to contain quite a lot of structure between the two surfaces. The

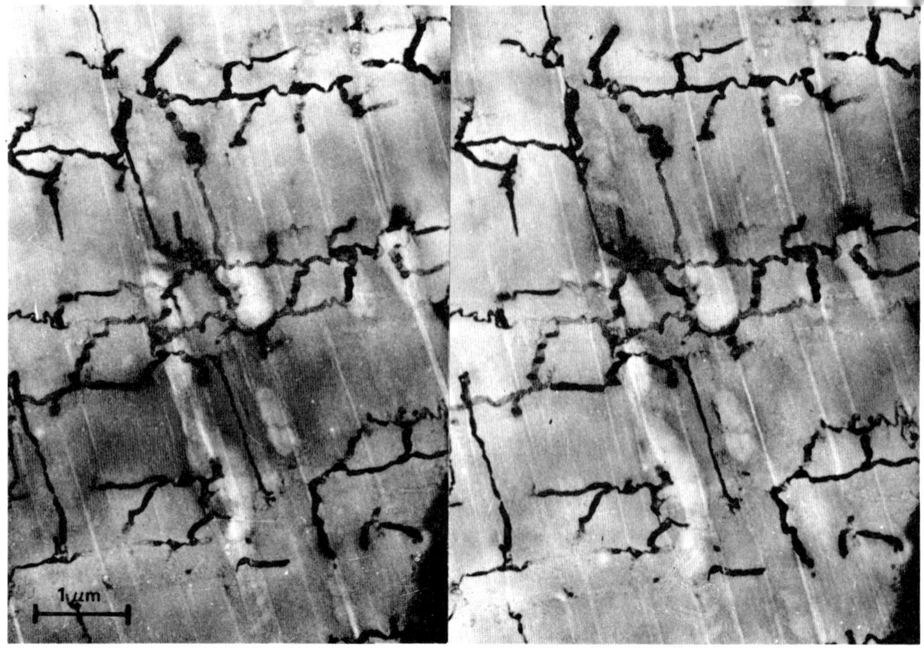

Figure 50 Stereo micrographs, showing the T-system of tubules in frog muscle. This section is 3 μm thick, and an accelerating voltage of 1000 kV was used. A method of seeing the stereo effect is described in the text.
(Courtesy of A. M. Glauert and L. D. Peachey; from A. M. Glauert (1979) Recent advances of high voltage electron microscopy. *Journal of Microscopy*, **117**, 93–101.)

problem with thick sections, of course, is that ordinary electrons cannot get through them. However, some electron microscopes have been developed to accelerate the electrons through a potential difference of up to 3000 kV. These electrons can penetrate sections several μm thick so that there are now no major difficulties in producing stereo micrographs.

Figure 50 is an example. Most people coming across stereo pairs like this one cannot use them because there is no viewer handy. However, no viewer is necessary if the following method is used. **(a)** The two micrographs are placed one beside the other, with the left-eye view on the right-hand side. In some cases, such as this one, it does not matter which way round they are. **(b)** Now you cross your eyes so that three images are seen, the middle one being the two micrographs on top of each other. At this stage, they will be out of focus. **(c)** The final step is to bring the middle image into focus. Once this is done, the stereo effect is quickly seen. I find this method extremely useful, but you should not persevere with it if you cannot make it work easily, in case it causes eye strain.

The stereo micrographs show the T-system of tubules in frog muscle. They are arranged in horizontal sheets, with the occasional tubule running from one sheet to the next. These tubules probably serve to transmit a stimulus from the nerve ending to all parts of the muscle fibre. The stimulus results in calcium ions being released from the endoplasmic reticulum into the cytoplasm. In turn, the calcium activates phosphorylase kinase, an enzyme which was mentioned earlier (figure 47b).

Apart from stereo micrographs, an advantage of the ultra-high-voltage electron microscopes is that they enable live cells to be examined. The cell, bacterium, or other specimen is placed in a small chamber containing gas and water vapour at normal pressure, and this is inserted into the microscope. The electrons are so penetrating that they go right through the chamber and the specimen, sometimes without killing it.

Scanning electron microscope

The scanning electron microscope (SEM) is another instrument that can produce vivid micrographs with a good sense of depth (figure 51). No stereoscopy is involved; instead, the sense of depth derives solely from the great depth of focus of the image. Every part from the nearest to the furthest away is focused. Photographers using close-up lenses know how difficult it is to achieve this with ordinary cameras.

When electrons strike the surface of an object, all sorts of things can happen. The electrons which arrive are known as **primary electrons**, and some of these can be scattered. At the same time, each primary electron can knock other **secondary electrons** out of atoms it collides with. Some of these secondary electrons will shoot off from the surface of the object. As we have already seen with the fluorescent screen (p. 90), some materials also give out light in response to electron bombardment. Yet another kind of emission is X-rays.

Most general-purpose SEMs form an image from the secondary electrons, though some can also give separate images using the X-rays or light. Perhaps the most remarkable thing about the SEM is that there are no lenses between the specimen and the image. Clearly the instrument is very versatile, and works on quite a different principle from ordinary electron microscopes.

There may be several demagnifying lenses between the electron gun and the specimen (figure 51b). Their function is to make the electron beam extremely narrow so that it forms a minute spot where it hits the specimen. At best this spot may be only 5 nm in diameter or even slightly smaller. The overall resolution of the SEM is determined by the smallness of the spot. Only between 1 and 30 kV are used for accelerating the electron beam, since lower voltages give more secondary electrons when the beam hits the specimen.

The specimen is tilted towards the electron collector. This attracts most of the secondary electrons, no matter in what direction they shoot away from the specimen; this is because they are not moving very fast,

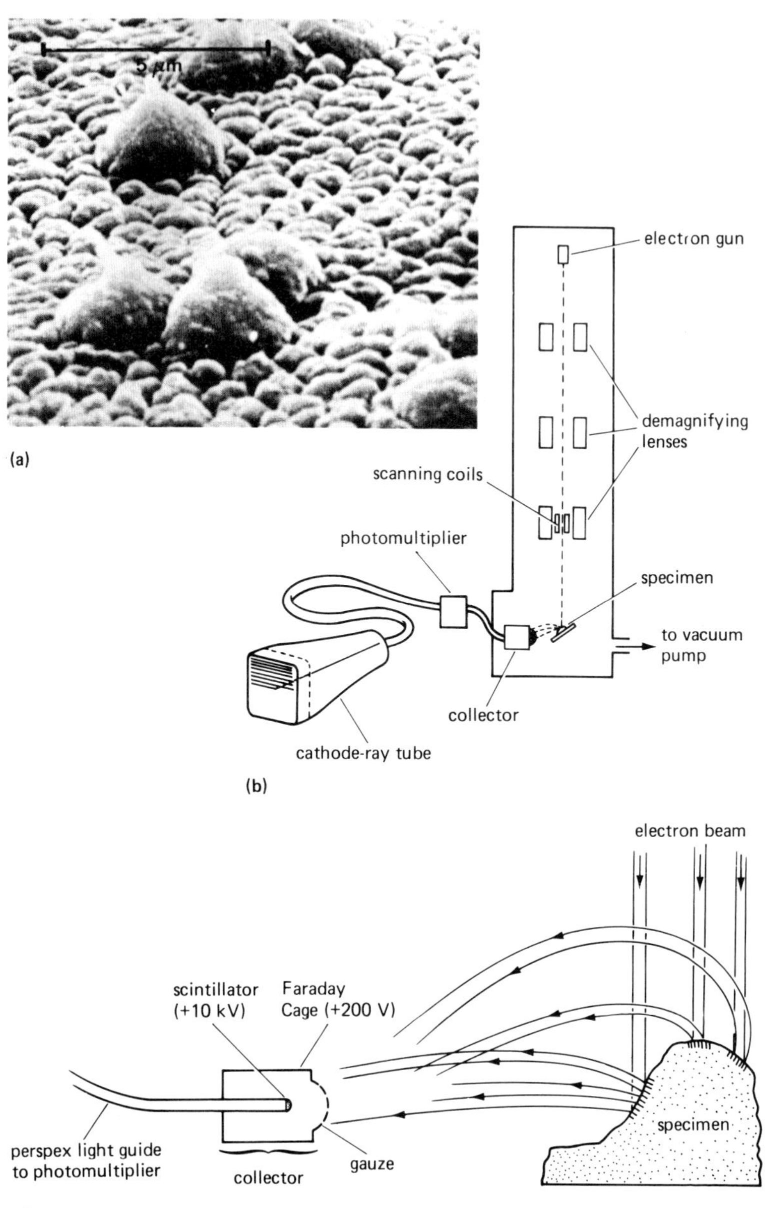

Figure 51 Scanning electron microscope. **(a)** Micrograph of sense organs on the surface of a parasitic platyhelminth. Note how the bumps *appear* to be lit from behind. **(b)** The general layout of the microscope, with details **(c)** of the way secondary electrons are attracted to the collector. In **(c)** the electron beam is drawn at three different positions to show how sloping surfaces give out more secondary electrons.

and are attracted by the positive charge of the Faraday Cage. The cage has a gauze cap at one end, and inside is a scintillator. When electrons strike the scintillator, tiny flashes of light are produced which pass along a transparent light guide to a photomultiplier. Both scintillator and photomultiplier are described in chapter 4 (p. 45).

So far it has been explained how the image from a 5 nm spot is obtained; but how is the picture of the whole specimen built up? This part of the mechanism is very like a television. The television picture is made up by a tiny spot of light scanning across the screen of a cathode-ray tube (see p. 134). Each scan makes a line on the fluorescent screen, dark in some places and light in others. The whole picture consists of hundreds of these lines, one above the other. The SEM image is produced in just the same way, using up to 1000 lines.

Exactly synchronised with the cathode-ray tube, the scanning coils inside the lowest demagnifying lens make the electron beam scan across the specimen. There is of course the same number of scans as in the cathode-ray tube. Parts of the specimen giving off a larger number of secondary electrons cause a brighter spot on the screen. This is such a different way of making an image from ordinary vision or transmission microscopes that it is not surprising that some strange effects are seen. In particular, the angle of a surface is largely responsible for the brightness of the image. Surfaces parallel to the plane of the photograph appear dark, while those sloping steeply away are generally brighter. An explanation for this can be seen in figure 51c; the electron beam illuminates a larger area of surface if the surface is sloping away from the electron source. So, a larger area is affected by electron bombardment, shown shaded, and this accounts for a larger emission of secondary electrons.

The SEM does not depend on transmission at all, so specimens of any thickness can be examined. On the other hand, since the secondary electrons escape only from atoms very close to the surface, the image gives information only about the surface. It does not reveal internal structure as the proper 3-D methods can.

Artefacts

An **artefact** is an effect produced by our treatment of the specimen; it was not there in the living organism.

Since artefacts are possible, some of the things we see on electron-micrographs may not really exist. If you consider how much we take for granted about the many organelles that are visible only with the electron microscope, then it would indeed be a calamity if we were just chasing illusions. Artefacts need to be taken seriously.

It will have been obvious, through this chapter and chapter 6, that preparing material for study sometimes involves remarkably violent actions. Some of the material may be destroyed altogether, such as washing away most of the cytoplasm from an ultrathin section. What remains may very likely have been altered to an unknown degree. It is easy to laugh at the eighteenth-century microscopists who imagined they could see the eyes and legs of *Amoeba*, but we may be doing the same kind of thing ourselves, however sophisticated we think our methods are.

Dr Harold Hillman of Surrey University has given much thought to the preparation of specimens. It is his conclusion that the trilaminar membrane, endoplasmic reticulum, Golgi body and pores in the nuclear membrane are all artefacts. They simply do not exist, or at least not in anything like the forms we are familiar with. He argues that any simple membrane will look trilaminar merely because the heavy atoms of the stain will collect along the two sides of it. The trilaminar appearance does not reflect the internal structure of the membrane. He argues that the frequent appearance of the endoplasmic reticulum as long, parallel lines cannot be a section of any real three-dimensional structure (figure 46b). Instead, it is the cytoplasmic solutes that form lines of precipitate during the preparation of the sections. There are arguments, too, for being suspicious of the Golgi body and nuclear pores.

Who, then, is correct: the biologist who believes everything he sees in an electronmicrograph, or the person who thinks most of it is artefact? Most likely neither. Having taken seriously the arguments of both sides, I believe the answer lies somewhere between the two extremes. There does seem to be something suspicious about, for instance, the endoplasmic reticulum. It is difficult to reconcile the kind of image seen in figure 46b with any three-dimensional structure of tubes or flattened vesicles. This is borne out by comparing the two micrographs in figure 46, for the striking differences seen in the reticulum as well as in other parts of the cell seem to be due to the fixative used. In short, the pattern obtained with one of the fixatives probably does not look much like the real reticulum.

As just hinted, one way to tackle the artefact problem is to compare the results of different methods of preparation. The more dissimilar the methods, the more convincing is any vindication of the micrographs. Freeze-etching is particularly important in this context. This is because it involves no fixing, embedding, sectioning, normal staining, or chemical dehydration. Yet nuclear pores, Golgi bodies, and many of the well-known organelles can still be easily recognised. This goes a long way towards convincing me that these structures are real, and would be seen, if it were possible to examine them, in living cells.

9 Measuring light

The last two chapters have described how to see with light and electron microscopes. What about seeing with our eyes? Our eyes are marvellous pieces of equipment, even when like mine they suffer from astigmatism, myopia, *and* colour blindness. To treat defects such as these, much research has been devoted to finding out how the eyes work.

Astigmatism and myopia result from a defective lens system in the eye. They can usually be corrected with spectacles or contact lenses. These help focus the light onto the retina at the back of the eye. But they cannot compensate for defects in the light-detecting cells: so they are no use for colour-blindness.

Colour vision

The measurement of light has played an important role in studying colour vision, but finding out how colour vision works is no easy matter. There are four pigments in the retina which are used for detecting light. One of these is rhodopsin, the pigment that absorbs light falling on the rod cells. Rods are far more numerous than the cone cells, and so it has been possible to extract rhodopsin and study it in the test tube. Accordingly, quite a lot is known about it. But rhodopsin is used for vision in low light intensities, such as at night. It is not used in conjunction with any other pigments, and this means that it can only detect light intensity. It cannot be used to distinguish colours.

All the other pigments are located in the cones. They are used for colour vision in bright light. Unfortunately, the cones are not very numerous and so it is difficult or impossible to extract the pigments for study in the test tube. How then can the biologist investigate them? This chapter attempts to explain.

Monochromatic light

Observing the colour of light can be just as important as noting its intensity. For instance, when a retina adapted to the dark is exposed to light, it goes through a series of colour changes. The original pinky-red colour is due to rhodopsin. As it breaks down to opsin and a form of vitamin A it changes to orange, and then to yellow before fading completely. These visible colour changes indicate three of the chemical processes occurring.

The colour of light is due to its wavelength. If light rays have only one wavelength, they are called 'monochromatic' and are of a pure colour. Thus light of wavelength 700 nm is pure red. But there is not much monochromatic light to be found, except where it is deliberately produced in a laboratory or a sodium street lamp. Most light is a mixture of wavelengths, so even if a biologist has good colour vision, it is not always easy to make objective observations of colour. Worse still, there is no way the naked eye can detect how many wavelengths are present and in what proportions. Describing light by the wavelengths and the quantity of each would be more objective and also give more information. Since this cannot be done by the naked eye, instruments are needed.

Glass lenses work by bending light. They collect light from a range of angles and focus it to a point (figure 37). As is mentioned in chapter 7 (p. 77), a problem with lenses is that the rays of different wavelength bend by different amounts. So, because white light consists of many different wavelengths, it is difficult to make them come to a focus at the same place. The red light focuses at one spot, and the blue light at another.

Although this chromatic aberration is a nuisance in a lens, it is very useful in a **prism** used for making monochromatic light. A prism is like a lens which has flat surfaces rather than curved ones. To make mono-

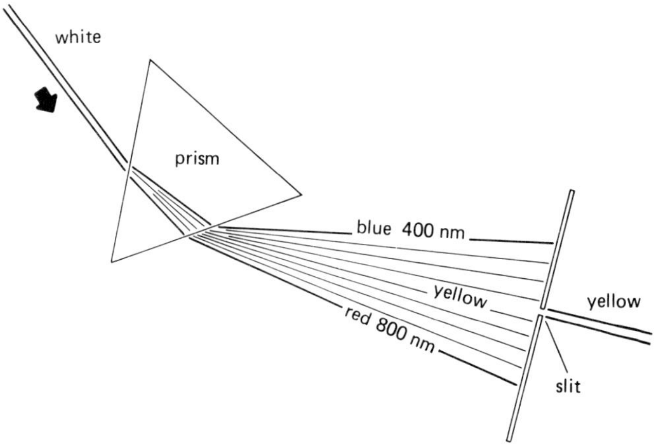

Figure 52 The spectrum. White light is split into its component colours by a prism. A slit is used to select a narrow range of wavelengths. The light passing through a very narrow slit is effectively monochromatic.

chromatic light, a thin beam of white light is directed at one surface, from an oblique angle. The light rays bend as they enter the glass, and they bend again as they leave it on the other side of the prism (figure 52). As a result the different colours spread out to form a spectrum. The visible spectrum includes wavelengths from about 400 nm to roughly twice that.

Where a prism is not convenient, a diffraction grating can be used. It works on a different principle, but produces the same spectrum. With either method, monochromatic light or light with a very narrow range of wavelengths can be selected from the spectrum just by blocking off all the unwanted light. This is like tuning a radio by selecting just one waveband. Obviously, a narrower slit gives light which is more nearly monochromatic, but at the same time less light gets through. So usually a very bright bulb is used as a light source.

Action spectra

Monochromatic light is used in two ways. In the simplest case each wavelength is shone onto an organism or cell, and the response to the light is measured. One must ensure, of course, that the light intensity is the same with each wavelength. Then, when this measurement is plotted against wavelength, the resulting graph is called an **action spectrum**.

Any kind of response or action of the living material can be measured. For example, colour vision has been intensively investigated by testing people's colour perception. In the tests, a spot of coloured light is shone onto a screen the rest of which is lit with another colour. Careful analysis of the results led to the theory that there are at least three different mechanisms for detecting bright light. The action spectra of these are shown as the top three curves in figure 53. The three unknown mechanisms were found to have maximum sensitivity to light of wavelengths 440, 540, and 580 nm respectively. The curves are derived solely from people's responses to the tests, and are not based on studies of pigments.

Colour-blindness is particularly useful in studies of this kind. This is because colour-blind persons are thought to be defective in one of the mechanisms, or even sometimes two. If two mechanisms are missing, the third one can only detect light intensity, not colour. This is total colour blindness. For colours to be separated, two or three mechanisms are needed. The reason for this seems to be that two colours are separated only if different numbers of mechanisms are stimulated, or if they are stimulated to different degrees.

It was said earlier that colour blindness is not corrected by lenses. However, it is possible to compensate for some kinds of colour-blindness. For instance, a person who is red–green colour-blind can look through a red filter with one eye, and a green filter with the other. The flowers of a

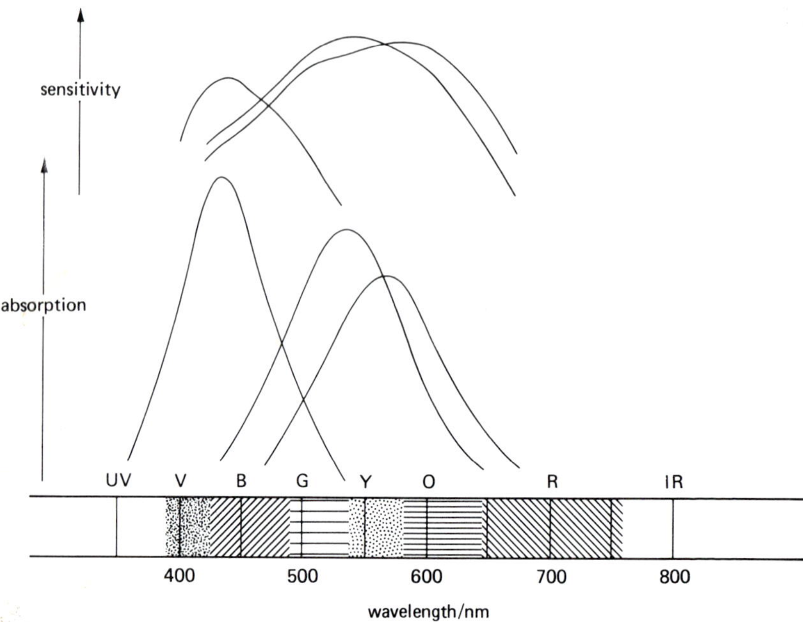

Figure 53 Colour vision. The upper three curves are the action spectra of three light-sensitive mechanisms, based on studies of colour perception. The lower three curves are absorption spectra for the 'blue-sensitive', 'green-sensitive', and 'red-sensitive' cone pigments. Under the horizontal axis are shown the colours in the spectrum: UV, ultraviolet; V, violet; B, blue; G, green; Y, yellow; O, orange; R, red; IR, infra-red. The visible colours are shaded.

red geranium might normally look the same colour as the green leaves. But although the flowers are red seen through the red filter, through the green filter they are black; then they really hit you in the eye!

Spectrophotometers

Instead of observing the response of a cell or organism, one can measure the absorption of light. This method is called **spectrophotometry**, literally 'measuring light across the spectrum'. It can be used on anything from a purified chemical compound to living tissue. The procedure is similar to measuring an action spectrum, except that a light-measuring instrument is provided to receive the light passing through the substance. An **absorption spectrum** is the final result, that is, a graph of absorption plotted against wavelength.

There are many designs of spectrophotometer, and a simple one is shown in figure 54a. Light is produced from a light bulb, and a thin beam is directed onto a prism or diffraction grating. By rotating the prism or grating, any chosen wavelength can be projected through the second slit. For good resolution between neighbouring wavelengths, the slit should be as narrow as possible. This beam, monochromatic or almost so, is usually 'chopped' by a mirror mounted on a rotating transparent disc. As it spins, it alternately moves into and out of the light beam. The light therefore passes alternately through the sample and through a control.

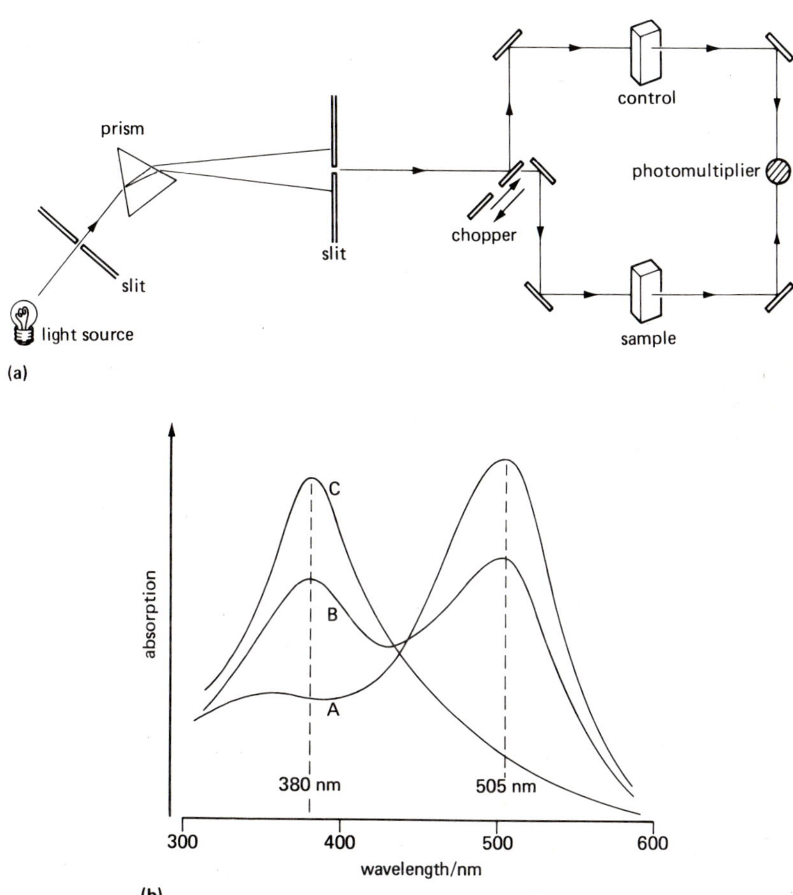

Figure 54 **(a)** Spectrophotometer. Light is separated into a spectrum by a prism or diffraction grating, rotation of which is used to select the wavelength going through the second slit. The 'chopper' is a mirror which is rapidly moved into and out of position so that light passes alternately through the sample and control. **(b)** Changes in the absorption spectrum as rhodopsin (A) is broken down via an intermediate stage (B) to opsin and retinal (C).

Usually the control is made up in the same way as the sample being investigated, except that the light-absorbing substance is missing.

The measurements needed are the light transmitted through the sample and the light that would have passed through it if the absorbing substance was not present. This second measurement is provided by the control, which is therefore essential. For each wavelength, the light passing through each container is measured by the photomultiplier (see p. 45), and from these measurements the absorption is arrived at. In this way, the light bulb's varying intensity of radiation is allowed for.

Spectrophotometers are useful partly because they are so versatile. Not only are they suitable for visible light, but with a little modification they can also cope with ultraviolet and infra-red. For ultraviolet radiation, any prisms, lenses, and containers it passes through must be made of quartz. For infra-red, a heated metal rod is substituted for the light bulb, and a thermocouple or some other detector for the photomultiplier.

Absorption spectra

Although the cone pigments are not easily isolated and purified, their absorption spectra can still be measured. This is done under the microscope, with the 'sample' beam being shone through a single cone cell, and the 'control' beam through the same cone cell after the pigment has been bleached by a bright light. It is found that the absorption spectra of cones fall into three groups, shown as the lower set of curves in figure 53. The simplest explanation is that there are three pigments, one in each of three kinds of cone. From left to right these are known as 'blue-sensitive', 'green-sensitive', and 'red-sensitive'. However, the labels should not be taken too literally; each pigment absorbs across a wide range of wavelengths, and the 'red-sensitive' pigment actually absorbs most strongly in the yellow. It is an interesting puzzle trying to work out what colours these pigments are themselves. For instance, one thing the 'blue-sensitive' pigment cannot be is, of course, blue.

The action spectra and absorption spectra in figure 53 are measured in entirely different ways. Although they are rather different shapes, one striking feature emerges from comparing the two sets of curves. This is that the maxima of each action spectrum and the corresponding absorption spectrum coincide almost exactly. This is good evidence that each of the three mechanisms relies on one of the pigments for detecting light.

You may already be familiar with the spectra of chlorophyll. The absorption and action spectra can be measured independently, but they show a striking similarity.

Absorption spectra have other uses. One of these may be illustrated by rhodopsin, the night-vision pigment. It was described earlier how the colour of the retina changes from pinky-red through orange and yellow before fading. This ought to be reflected by changes in the absorption spectrum. Rhodopsin absorbs light most strongly at a wavelength of 505 nm. It breaks down to opsin and retinal, an oxidised form of vitamin A. It is the retinal which is coloured at this stage, and it absorbs most strongly at 380 nm.

Now, 380 nm is about the lower limit of the visible spectrum, and over most of the spectrum retinal does not absorb strongly. Therefore, it is not obviously coloured. Clearly the naked eye is not much use for making exact measurements in this situation. The great value of the absorption spectra can be seen in figure 54b. In the course of the chemical reactions, the shape of the curve completely changes. It is therefore easy to measure the rate at which the reactions occur, simply by following the changes in absorption at a carefully selected wavelength such as 505 nm.

Oxidative phosphorylation

Methods such as the above played an important role in the investigation of oxidative phosphorylation. An early development was the construction of an action spectrum in yeast. The wavelength of light affects the rate of oxygen consumption, and the resulting spectrum resembles figure 55a. A notable feature is the three peaks. Although little was known at that time about the substance or substances responsible for the spectrum, there was a striking similarity with the absorption spectrum of oxidised haemoglobin. The three peaks in haemoglobin, called alpha, beta and gamma, are characteristic of compounds containing haem. This is the flat, plate-like structure in the haemoglobin molecule which includes the iron and carries the oxygen. It therefore seemed that the substance in yeast might contain haem.

Figure 55c shows what haemoglobin looks like in a simple spectrophotometer which displays the whole spectrum at the same time. This instrument is called a **spectroscope**, and although it gives a good qualitative idea of the spectrum, it is not really useful for measuring. As haemoglobin becomes deoxidised, the alpha and beta bands fade, so that it is easy to watch the chemical change taking place.

In the 1920s, Keilin was using this instrument to follow what was happening in living muscle (figure 55d). Unlike haemoglobin, the substances did not have many absorption bands when well oxygenated. But as the oxygen in the tissue was used up, at least six bands appeared in the spectrum. Clearly this was not haemoglobin, but the resemblance was closer than it might at first have seemed. Keilin argued that three compounds were involved, each having one alpha and one beta band, so they probably contained haem. It is now accepted that they are three

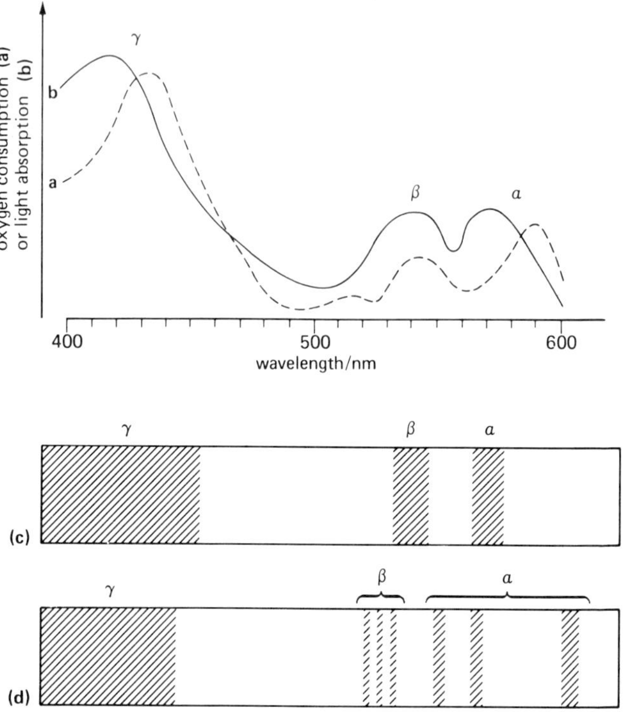

Figure 55 Respiration. **(a)** Action spectrum of cytochrome oxidase in yeast, and **(b)** absorption spectrum of oxidised haemoglobin. The essential features of **(b)** are seen in a spectroscope which displays the whole spectrum **(c)**. **(d)** shows the complex spectrum observed by Keilin in insect muscle. Bands only appear as the oxygen is depleted.

different kinds of cytochrome, one identical with the substance, cytochrome oxidase, which is responsible for the action spectrum in yeast.

When the absorption spectra of all the cytochromes are taken together, it begins to get complicated. To these must be added the spectra of NAD^+, $NADP^+$, and FAD. However, by careful study of the composite spectrum, it is possible to select a couple of wavelengths to represent each of the compounds. By following the change in absorption at these wavelengths, the oxidation stage of each compound can be traced without too much interference from the other compounds. In this way it has been possible to determine the order in which the hydrogen and electron carriers start working during oxidative phosphorylation. These measurements on living tissues have confirmed the conclusions reached from studies on isolated parts of the full sequence. A good description is given by J. A. Ramsay in his *Experimental Basis of Modern Biology*.

Identifying substances

Apart from the other uses already mentioned, spectrophotometry can be used to identify substances. The absorption spectra of many compounds have been published, especially in the infra-red. If two compounds have identical spectra they are almost certainly the same molecule, or very similar. So comparison of spectra can be useful for identification, rather as keys are used for naming species of organisms.

More than this, the structure of unknown compounds can be worked out to some extent. We have just seen an example in the case of the cytochromes, for the alpha, beta and gamma bands suggested the presence of haem. This approach can be taken much further. Chemical bonds such as $C=O$, $C=C$, $O-H$ and $C-H$ have characteristic positions in an ultraviolet absorption spectrum. Infra-red spectra are probably even more useful in this respect. Often all the absorption bands can be accounted for in terms of a particular chemical bond being caused to move in a particular way by the radiation. The five standard forms of movement are called stretching, rocking, wagging, twisting and rotating. Thus, the stretching of an oxygen–hydrogen bond can cause strong absorption at about 3000 nm wavelength; so a band at that position often indicates a hydroxyl group. An experienced infra-red spectroscopist can look at the spectrum of an unknown compound and say quite a lot about how the molecule is constructed.

Colorimetry

In some investigations, a complete absorption spectrum covering a wide range of wavelengths can provide much information, but later stages of the same study may not need the full spectrum to be measured each time. Instead, it might be more important to make a single measurement continuously, perhaps to follow a fast chemical reaction. For instance, in figure 54b, the decay of rhodopsin could be followed simply by measuring the change in absorption somewhere near 505 nm.

The simplest **colorimeter** (figure 56) uses a filter to select a band of wavelengths from the light emitted by a bulb. At the narrowest this band will still include wavelengths spanning about 100 nm. If the absorption spectrum of the coloured substance is not known, the filter should be chosen which makes the solution appear to absorb light most strongly. This means the filter is transmitting light which coincides with a maximum on the full absorption spectrum.

Perhaps the most frequent function of a colorimeter is to measure the concentration of a solution. The most reliable method is to make a set of solutions of the substance you are interested in. These solutions have different concentrations. Each one is placed in the colorimeter and its light absorption is measured. From the data a graph can be plotted of the absorption against concentration. Then the concentration of any solution of unknown strength can be found by interpolating its absorption in the graph.

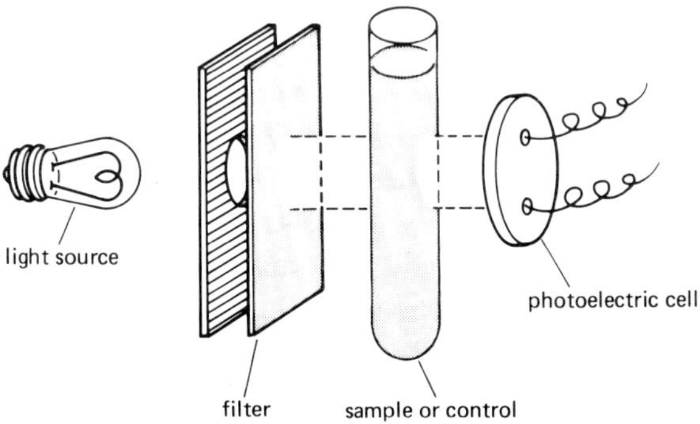

Figure 56 Colorimetry. In a simple colorimeter, a band of wavelengths is transmitted by a filter (or a prism and a slit). A photoelectric cell measures light transmitted by the solution and the control.

10 X-rays

Various kinds of electromagnetic radiation have been mentioned earlier in the book. Gamma rays appeared in chapter 4 as a radiation emitted from unstable isotopes. Visible light was the subject of chapter 7. In addition, in the last chapter the absorption spectra of infra-red and ultra-violet rays were shown to be useful. All these kinds of radiation are part of a continuous spectrum (figure 57).

Cosmic and radio waves are not of much interest to the general biologist. On the other hand, X-rays do have a number of special uses. There are X-ray microscopes, though of a rather peculiar kind. X-rays enable us to see mineral objects, such as bones and swallowed safety pins, inside living organisms. The complex method of Fourier synthesis enables the detailed structure of macromolecules such as proteins to be worked out. And probably most famous of all, X-rays provided some of the most important original evidence for the double-helix theory of DNA.

X-ray microscopes

Perhaps surprisingly, one thing X-rays have not been useful for is building high-resolution microscopes. In chapter 7 (p. 75), it is explained how wavelength fixes the minimum size of Airy's disc, and this in turn determines the resolution of a microscope. Ultraviolet radiation enables a slight improvement to be made in comparison with visible light; so why do X-rays not allow even smaller things to be seen?

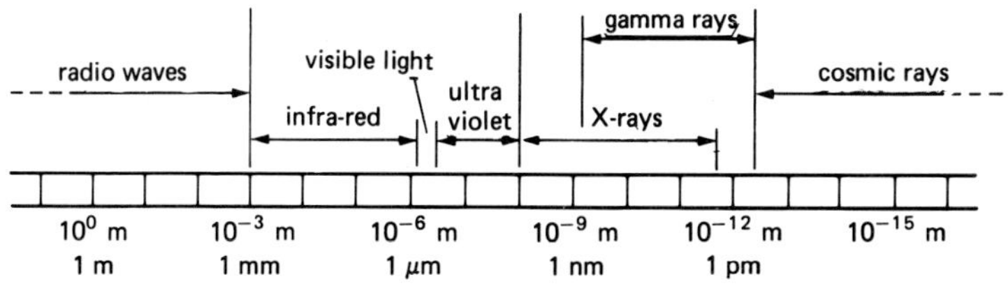

Figure 57 The electromagnetic spectrum. The scale of wavelengths is logarithmic. UV: ultraviolet.

The problem is in trying to focus X-rays. They are hardly bent by passing through glass or other materials, so ordinary lenses are of no use. Not consisting of charged particles like electrons, they do not respond to electric or magnetic fields either; so electromagnetic lenses are also useless. Sometimes mirrors can do the work of a lens, but X-rays are absorbed and not reflected by a mirror.

However, **X-ray microscopes** do exist. As in the scanning electron microscope, they have no lens between the specimen and the image. Instead the X-ray instrument is a 'point projection' microscope. In principle this is no different from holding your hand near a small light bulb, and producing a large shadow of the hand on the wall. The magnification is equal to the ratio of the distances from bulb to wall and from bulb to hand.

To achieve a crisply focused image, the source of radiation must be much smaller than the specimen. Indeed, it must be smaller than the resolving power. To resolve two points 1 μm apart, the source of X-rays must be a point smaller than 1 μm in diameter. This explains the 'point' in point projection. The main reason why good light microscopes cannot be built on the point-projection principle is the difficulty in making a small enough light source of sufficient intensity. For similar reasons, the X-ray microscope cannot achieve a really good resolution. X-ray microscopes are, however, useful for examining insects or other organisms which are opaque to light.

Diffraction

It would be a pity if the short wavelengths of X-rays could not be used somehow to achieve a fine resolution, just as has been done in the electron microscope. Fortunately, the methods of X-ray diffraction can do this, though not by forming an image directly. Any image, of a molecule for instance, that does finally emerge is the result of extensive calculations based on a complex theory. Only the barest sketch of the methods can be attempted here.

You do not need to understand the theory of **diffraction** in order to appreciate its fascination and beauty. If you have access to a pair of binoculars, you can try this experiment. When it is dark, find a distant light, as far away as possible, and look at it through the binoculars. The binoculars will merely produce an enlarged image of the light; but if you now watch the light through a net curtain, a very different effect is seen. A pattern of spots is superimposed on the area around the light (figure 58a). It is obviously not an image of the curtain, because it does not move when you move your head. On the other hand, it is only there when the curtain or something similar is present.

It is worth trying other fabrics. I have a drip-dry nylon shirt which gives a very different pattern compared with that of the curtain. Further variations result from changing the light source. For instance, white car headlamps cause multicoloured patterns, whereas a sodium street lamp gives simpler patterns of a single colour. This is because the sodium lamps produce light which is fairly monochromatic (see p. 106).

These are diffraction patterns produced by the holes in the fabric. The X-ray crystallographer has to interpret similar patterns resulting when X-rays are shone through a crystal. In both cases, the fundamental cause is the same. When light passes through a curtain, each hole acts as though it were a light source in its own right, sending new light waves out in all directions. Similarly, when X-rays strike a crystal, each atom

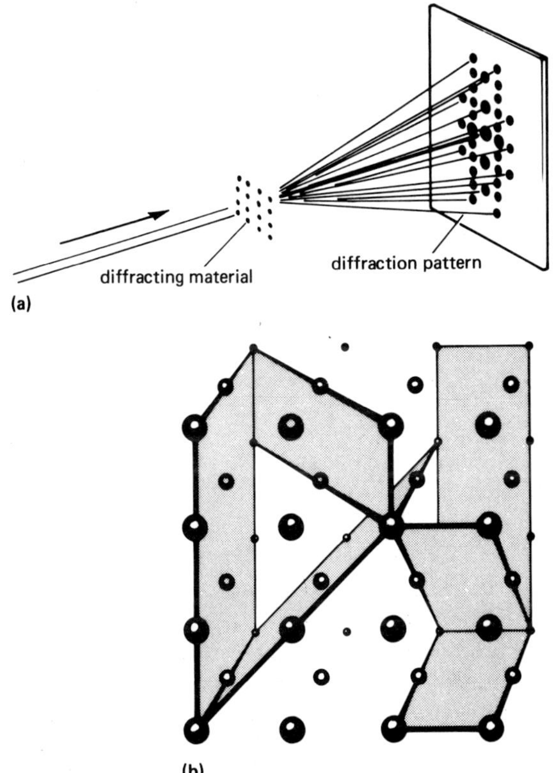

Figure 58 Diffraction. **(a)** When a beam of light or X-rays hits a diffracting material such as a curtain or a crystal, some passes straight through to form the central spot of the pattern. The other spots appear only where diffracted rays arrive at the screen in phase. **(b)** A few of the lattice planes in a crystal which has atoms at the corners of a cubic unit cell: only a small portion of each lattice plane is shown.

emits new X-rays in all directions. The curtain and the crystal are acting like a battery of tiny radiation emitters, lined up in regular rows and columns like the trees in an orchard.

So why is the screen in figure 58a not uniformly bathed in this scattered light or X-radiation? The answer lies in the phenomenon of interference. This is described elsewhere in connection with phase-contrast microscopes (p. 85). Suppose the radiation emitters in the diffracting material are atoms in a crystal. Now, over most of the screen, the waves from some of the atoms will arrive out of phase with waves from other atoms; so these will cancel each other out, and the screen will not be illuminated by X-rays. However, at certain points, all the waves from the different atoms will be arriving in phase. It is only at these points that the screen is illuminated. If a film is used as the screen, a black patch appears on the photograph at each illuminated point.

In a crystal, the situation may be complicated. After all, there are lines of atoms running at many different angles in three dimensions. Calculating where the spots are to be expected in the diffraction pattern might seem an almost insuperable task. However, an important discovery was made by Sir Lawrence Bragg; he found that each spot is produced exactly as though it had been reflected like light from a lattice plane in the crystal. A **lattice plane** is simply a flat sheet of atoms running right through a crystal. Figure 58b shows a small portion of a crystal; only six of the lattice planes have been indicated, and you can probably see at least another six that could have been drawn instead. A single diffraction diagram may consist of hundreds of spots, and each one is a 'reflection' off one set of lattice planes.

Fourier synthesis

One diffraction pattern differs from another in the positions and intensities of the spots that make it up. These spots are a kind of coded information. In fact, they contain all the information needed to make a picture of the unit cell of the crystal. The method of doing this is called Fourier synthesis. In figure 58b, the **unit cell** is the structure that makes up the crystal simply by repeating itself; it is a cube with an atom at one corner. If a protein or other macromolecule can be crystallised, the unit cell may contain just one protein molecule.

The parts of the protein molecule that scatter the X-rays are the electron clouds of the atoms, so the picture of the protein that results from a Fourier synthesis is a map of the electron clouds. Heavy-metal atoms, with their dense clouds of electrons, show up clearly; at the other extreme, hydrogen atoms hardly appear at all. But with adequate resolution, the map of electron density gives a good idea of the molecule's structure.

Figure 59 is a map of the electron density of a pyrimidine molecule. Pyrimidines are one of the two kinds of base found in nucleic acids. A ring of six atoms is the main feature, and in this particular example the two side chains on the right-hand side are $-NH_2$ groups. At the other end of the molecule is a substituted chlorine atom. A protein molecule has, of course, a much more complicated map of electron density, and working it out in three dimensions takes an immense amount of calculation. But then an impressively complete knowledge of the protein's structure is achieved, including the way the amino acid chains are coiled and twisted around each other.

A commonly used type of X-ray has a wavelength of 0.15 nm, so one hopes for a resolution of about 0.08 nm. This can in fact be achieved, and it gives a picture slightly more detailed than figure 59, which represents a resolution of 0.12 nm. However, if the resolution is only slightly worse, the picture becomes unrecognisable and so rather useless, at least for small molecules like pyrimidines.

The way the resolution is controlled is of some interest. If only the spots near the centre of a diffraction diagram are used, a poor resolution is the result. As dots further away from the centre are added in to the calculations, so the resolution improves. For the best resolution all the spots need to be used. Therefore, the central spots give only a rough image, and the outer ones add the fine detail. This is just like the light microscope. The reason for wanting an objective with high numerical aperture (figure 40b) is that it captures more of the widely diverging rays. These contain the information needed for a really detailed picture of the specimen.

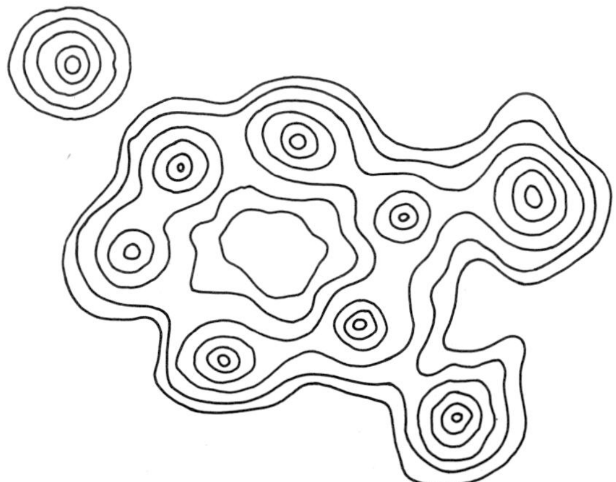

Figure 59 Fourier synthesis. The electron-density map of a chlorine-substituted pyrimidine molecule. Apart from the six-atom ring, there are two amino groups at the right and one chlorine atom at top left. (0.12 nm resolution.)

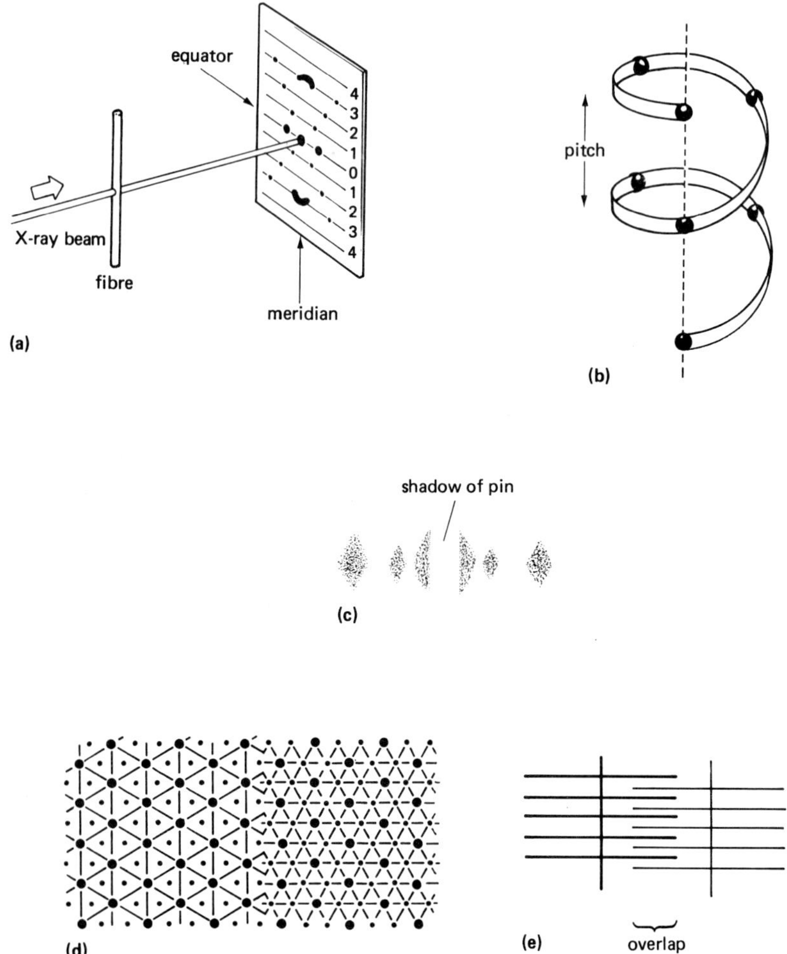

Figure 60 Fibre diffraction. **(a)** Spots on a level with the centre of the pattern are equatorial' and those on the vertical axis are 'meridional'. Any others are 'diagonal'. Layer lines are numbered 1 to 4. **(b)** Model of a helical molecule in which there are three chemical repeat units in each turn. **(c)** X-ray diffraction from striated muscle. As the muscle contracts or relaxes, one pair of equatorial spots decreases while the other increases in intensity. The central spot is largely missing because it is scattered by a metal pin. **(d)** A cross-section of a muscle showing the arrangement of thin actin and thick myosin filaments. The outer 'reflections' in **(c)** are caused by the lattice planes shown on the left, and the inner by those on the right. **(e)** The pattern in **(d)** is only found in the region of the sarcomere, where the thin actin filaments overlap the thick myosin filaments.

Fibre diffraction

X-ray diffraction can also be useful with biological materials that do not form proper crystals. So long as the material contains suitable lattice planes of atoms or other components, it can 'reflect' X-rays and produce a diffraction diagram. This method can even be used with living tissues. Muscle is one example.

Fibre diffraction patterns are obtained by shining an X-ray beam at a vertical fibre of biological material. This can be a living muscle fibre, but equally it can be silk, polysaccharide, nucleic acid, or some other **paracrystalline** material. In a paracrystalline fibre, there are many elongated units such as molecules arranged parallel to each other along the fibre axis, but otherwise they are disordered. Therefore, the fibre is not made up of a regularly repeated unit cell; hence it is not properly crystalline and a Fourier synthesis cannot be carried out.

Nevertheless, much useful information is provided by a diffraction diagram. Any equatorial spots (figure 60a) are 'reflections' from vertical lattice planes, just as a horizontal light beam striking a vertical mirror can only be reflected horizontally. Some of the strongest 'reflections' from muscle fibres are equatorial spots, usually two on each side of the centre. As the muscle elongates after a contraction, the outer spots are seen to become less intense while the inner ones do the opposite (figure 60c).

This can be explained by the well-known pattern in cross-sections of striated muscle. There are two sets of lattice planes, shown in figure 60d. Both exist in the region where actin and myosin overlap. However, the left-hand set includes only myosin, and so this is found also where the myosin is not overlapping the actin. Therefore, when the muscle is elongated, the 'reflection' from the left-hand set of lattice planes is strong, while that from the other set is weak because of the short length of overlap. As the muscle contracts, the 'reflection' from the right-hand lattice planes becomes more intense as the amount of overlap increases.

Muscle gives much more complicated diffraction patterns than this. If the very strong equatorial spots are ignored, much weaker meridional 'reflections' can be detected. These give information about the spacing of components up and down the fibre, such as the cross-bridges linking the actin and myosin.

DNA

Fibre diffraction has been particularly valuable for studying helical structures. In the case of a helical or spiral molecule like that in figure 60b, several of the important dimensions can be worked out from the diffraction diagram. Many of the diagonal and meridional spots are arranged on regular layer lines, above and below the central spot. The vertical distance from the equator to the first layer line gives the pitch of the helix, that is, the distance along the fibre occupied by one turn.

A fundamental principle of diffraction diagrams is that measurements vary inversely with the corresponding dimension in the molecule. A smaller measurement on the diagram means a larger one in the real structure; and vice versa. Now, another dimension of interest is the distance along the fibre axis from one chemical repeat unit to the next. These chemical repeat units would be sugars in a polysaccharide, amino acids in a protein, and nucleotides in a nucleic acid. All these can form helical molecules. In figure 60b, there are three repeat units per turn, so the distance from one to the next is one third of the pitch. Being smaller than the pitch, the spot representing this distance will be found further away from the centre of the diffraction pattern than the first layer line. In fact, the spot is meridional, and in figure 60a it is seen on the third layer line. The number of the layer line gives the number of chemical repeat units per turn of the helix.

When Watson and Crick were investigating the structure of DNA in the early 1950s, they were very keen to see X-ray diffraction diagrams of the substance. They themselves spent most of their time building models and doing calculations to evaluate them; each model was a hypothesis. They knew about Chargaff's experimental evidence derived from separating the nucleotides by ion-exchange chromatography (see p. 58). He had found that the adenine and thymine molecules were equal in number, and so were the guanine and cytosine. But Watson and Crick needed observations that would bear directly on the shape of the DNA molecule. X-ray diffraction offered the best hope.

Unfortunately, neither Watson nor Crick was an X-ray crystallographer. They were dependent on seeing the results obtained by Rosalind Franklin and Maurice Wilkins in London. But it was none too easy getting hold of the photographs! And all the while a rival chemist, Linus Pauling, was trying to solve the same problem in the USA, and might beat them to it. For the exciting story of this famous piece of research, you should read Watson's book *The Double Helix*.

The first clear diffraction patterns to be obtained were disappointing. They are now referred to as the 'A' form of DNA; the molecules are fairly crystalline and give complicated diagrams not obviously showing a helical structure. But later, a different pattern (figure 61) was produced from DNA containing more water. This is now known as the 'B' form. It was the evidence Crick and Watson were hoping for.

One obvious feature is the strong though rather diffuse 'reflection' on the tenth layer line. This indicated that something was repeating itself every 0.34 nm along the fibre axis. It was known that the flat base molecules are about that thickness, so they seemed to be piled on top of each other like a pile of coins, each at right angles to the fibre axis.

Another striking feature is an absence of meridional 'reflections' in most of the diagram, and instead a prominent diagonal row of spots. This suggested a helical molecule, because few other regular arrangements of atoms could produce such a pattern. Most of the diffraction is caused by the phosphorus atoms, since they have denser electron clouds than any of the other atoms in DNA. Viewed from the side, the phosphorus atoms appear to form almost straight lines running in two directions. These can act like lattice planes, 'reflecting' X-rays in a direction at right angles to them. So the angle *a* in the pattern indicates the angle α the helix makes with the horizontal.

Given that the diffraction came from a helix, the distance from equator to first layer line indicated a pitch of 3.4 nm. From the position of the meridional patches on the tenth layer line, it seemed that there were ten chemical repeat units in every turn of a helix.

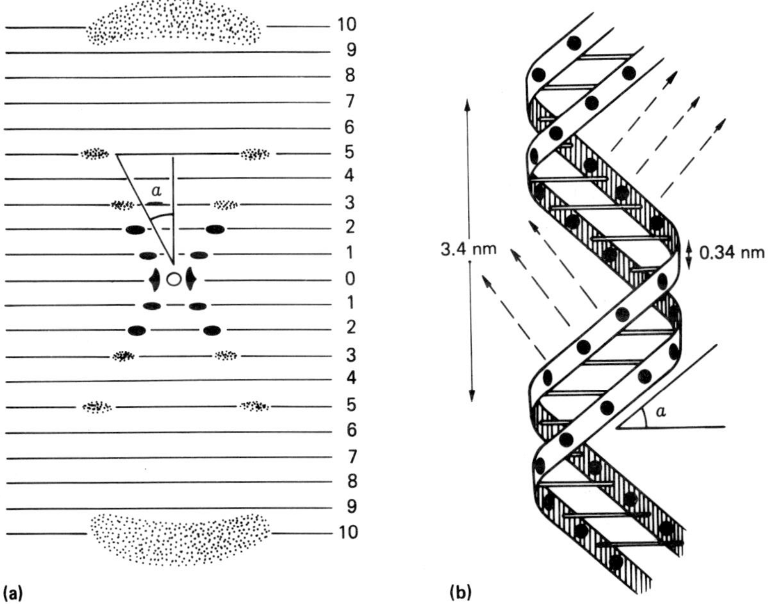

Figure 61 DNA **(a)** X-ray diffraction pattern from the 'B' form of DNA. **(b)** The double-helix model for which the X-ray pattern provides evidence. Each rung of the ladder is a pair of bases, adenine and thymine, or guanine and cytosine. The positions of the phosphorus atoms are indicated by black discs. Interrupted arrows show the direction of X-rays 'reflected' by the rows of phosphorus atoms.

All this gave a great boost to the morale of Watson and Crick, and guided them towards a helical theory for the DNA. What it did not immediately answer was the question: how many helices are there twined round each other? This had to be settled largely by building models and seeing which made most sense. In retrospect, though, an explanation could even be found for the absence of spots on the fourth and sixth layer lines. This was thought to be a consequence of having two helices separated from each other by about three-eighths of the pitch.

Important though the X-ray diffraction pattern was, the double-helix structure could not have been worked out from that alone. It functioned more as a way of evaluating hypotheses. Given Chargaff's evidence, and knowledge of the chemical building blocks of DNA, Crick and Watson selected a few hypotheses to work on. The X-ray evidence helped them discard some of these, so that in the end they were left with only one hypothesis which seemed to fit all the facts. This is the double-helix theory which has been the foundation of so many dramatic advances in biology over the last 30 years.

11 Electrical measurement

In this age of computers and microchips, it is amazing how many measurements do not need electricity at all. Chromatography, manometry and the light microscope can be used without it. One could measure length, mass, temperature, pH, oxygen concentration and osmotic pressure during a power cut if one had to. But a glance back through earlier chapters will show that electrical instruments are indispensable in many areas of biology. Also, variables like temperature and pH can often be measured more sensitively or conveniently with an electric meter.

Transducers

It is suggested in chapter 3 that, to understand an instrument, one needs to see how information passes through it. In the case of many electrical instruments, this is difficult unless you are adept at reading complex circuit diagrams. Therefore, no attempt is made here to trace the complete route taken by information through any instrument. We shall merely be concerned with the basic principles.

A biological phenomenon under scrutiny may or may not be electrical to start with. The charge on the surface of a nerve cell is; the temperature of the cell is not. But obviously the thing being measured must be converted into an electrical form in order for the instrument to deal with it. Anything which does this is called an **input transducer**. It is the means for feeding information into the instrument.

Input transducers are many and various. Here we shall look at only a few, used for measuring electrical potential, temperature and redox potential. Once the input transducer has produced a pattern of electrical signals in the instrument, the signals need to be converted once more into some other form. This is because we cannot detect electric currents or voltages with our own senses. The components which do this conversion are called **output transducers**. Two sorts are described later in the chapter: the ordinary meter for measuring current or voltage; and the cathode-ray oscilloscope.

In some cases, all we need to do is connect an input transducer to an output transducer, and, hey presto! we have a complete instrument. This is possible with some light meters, because they generate their own power. But something as simple as this is an exception. Normally there has to be at least an external power supply. **Amplification** is also frequently necessary. Take, for instance, the voltage difference between the inside and outside of a nerve cell. It may be about 0.05 V. But if the measurement is to be displayed on a cathode-ray oscilloscope, it must be converted to something like 100 V. In this case, the voltage must be amplified 2000 times. Sometimes, a current needs amplifying. An example is the scintillation counter (p. 45); here, a photomultiplier converts a minute trickle of electrons into a current 1 000 000 times stronger. Without amplification it could not be measured.

Two other things need to be done between the input and output transducers. The mains supply gives an alternating current, whereas many transducers work with a direct current, flowing in only one direction. Therefore, one form of current may need to be converted to the other. Calculations are also carried out in some instruments: the spectrophotometer in figure 54a receives signals from the sample and the control, and then works out the ratio between the two signals.

If all these processes are going on between the input and output transducers, it can be understood why complex circuitry is often needed. Readers keen on electronics may find it well worthwhile learning more about the circuits, but many biologists, myself included, find electronics rather baffling. So long as we have a general understanding of the transducers, we can regard the rest of the instrument as a 'black box' (p. 34). This is what we shall do here.

Potential, current, resistance

Input transducers can feed a signal into an instrument in three ways. A potential difference or current can be generated, or a resistance can be changed.

As an example of potential difference, let us return to the nerve cell. After making a measurement, we may write in a notebook: 'neurone resting potential = – 60 mV'. What does this mean?

It is well known that there is a sodium pump in the plasma membrane of a nerve cell or neurone. This pumps sodium ions out of the cell so that they accumulate just outside the plasma membrane (figure 62). Each sodium ion carries one positive charge; that is, it has one fewer electron than it needs to be a neutral atom. The electron has been left inside the cell, so just inside the plasma membrane there is an excess of negatively charged electrons.

Where negative and positive charges are equal in number, there is said to be zero **potential**. But inside the neurone plasma membrane there are excess electrons (a negative potential); while outside there is a deficit of electrons and so an excess positive charge (a positive potential). This

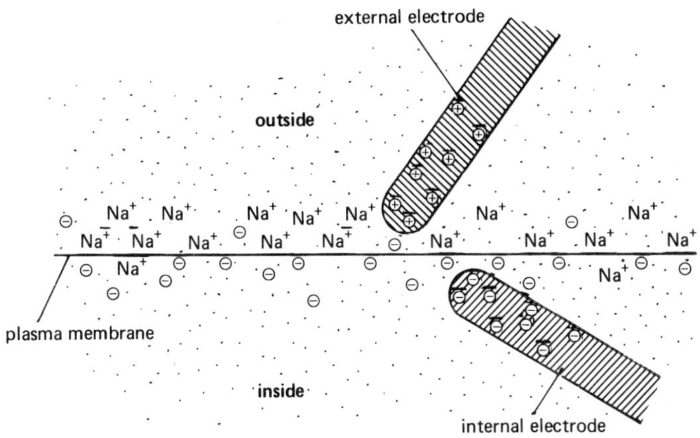

Figure 62 Resting potential at the plasma membrane of a neurone. Each electrode is given the same potential as the fluid it is immersed in. The minus and plus signs represent negative and positive charges.

means that there is a 'potential difference' across the membrane. It is measured in millivolts. The greater a potential, the greater is the concentration of excess positive or negative charges.

The neurone's potential difference is measured with a pair of electrodes. These are minute pieces of wire. One is inserted into the cell, while the other touches the outer surface. If the two electrodes are joined by an electric circuit, a **current** can be expected to flow through it. An electric current is simply a flow of electrons, whether along a wire, through a photomultiplier, or down an electron microscope column. There will be a current in this case because electrons repel each other. Where the potential is negative, just inside the plasma membrane, some of the electrons escape by entering the internal electrode. Meanwhile, just outside the membrane, there is a positive potential. Here, electrons will be attracted out of the external electrode by the positive charge. Adding electrons to one electrode, and withdrawing them from the other, makes a current flow through the circuit.

This current would probably flow for only a very short time, so it would be difficult to measure. There is another drawback to the method; the potential difference across the membrane would be destroyed. Measuring something and destroying it at the same time is not a good idea. So, what we need is some way of measuring the potential difference without allowing a current to flow.

In principle it can be done very simply. The way to stop electrons entering the internal electrode is to push electrons into it from the electric circuit. In other words the potential of the electrode is artificially made more negative until it has the same value as the cytoplasm. Then there is no electron flow into or out of it. Similarly, the external electrode has electrons withdrawn until it is at a positive potential equal to the fluid in which it is immersed. The observer knows when the electrode potentials are right, because then no current flows through the circuit from one electrode to the other. It remains to measure the potential difference between the electrodes, and this gives no problems. It is usually about 60 mV.

Having described potential and current, we are left with resistance. Electric cables are covered with a plastic insulator. and the whole point of the insulator is to prevent electrons escaping from the cable. The insulator is said to have a high **resistance** to their flow. But if we want a current, a metal wire is usually used to provide a route. Currents flow easily through it, because most metals have a very low resistance. The resistance of an object can therefore be measured by using it to join two points at different potentials. The terminals of a car battery would do: their potential difference is always close to 12 V, so long as the battery is not flat. Then the current flowing through the object is measured. The greater the current, the lower is the resistance.

Temperature

Resistance is used in some transducers, and one for measuring blood pressure is described in the next chapter (p. 140). Another is the **thermistor**, which measures temperature. The resistance of many substances increases with temperature, though carbon and some metal oxides do the reverse. Thermistors are usually made from oxides of nickel, iron, copper or other metal. They can be made very small indeed, so small that one can be mounted in a hypodermic needle and inserted deep into living tissue. The resistance many decrease by 50 per cent for every 20 °C temperature rise. In other words, twice the current will flow through it, provided that the potential difference between the two ends of the thermistor is kept constant. However, there is a problem; passing a current through it generates heat, altering its temperature, so the current must be kept as low as possible.

A very different kind of transducer for measuring temperature is the **thermocouple**. The principle was discovered as long ago as 1823. Two lengths of wire are needed, made of different metals. The end of one is joined to the end of the other, and the remaining free ends are also joined. This gives a loop with two junctions along it. If the whole thing is kept at the same temperature nothing happens, but if one junction is at a different temperature from the other, a current flows around the loop. The size of the current depends on the temperature difference. To use it for measuring temperatures, one junction is kept at a known temperature, in contact with melting ice, for instance. Then the size of current indicates the temperature of the other junction.

Incidentally, if a power supply is inserted into the loop to force a current around the opposite way, the loop absorbs heat at one junction and emits it at the other. This can be used to build a mini-refrigerator.

These electrical transducers have several advantages over mercury thermometers. Apart from being more sensitive, they can be placed at a considerable distance from the observer. Imagine the problems of measuring the temperature of the air in a forest canopy 50 m above one's head. Using a telescope to observe a thermometer on a tall pole is not really very satisfactory, but a thermistor or one junction of a thermocouple can be raised up on a pole or a balloon, leaving the output transducer next to the observer on the ground.

The small size of the thermistor or thermocouple junction also means that the temperature of tiny spaces can be measured. An ordinary thermometer with its large bulb might be quite useless. It is easy to see which instruments are better for studying the temperature of, say, a spider's brain.

Yet another convenience, which is true of all electrical instruments, is that a variety of output transducers can be used. For ordinary purposes, a thermistor or thermocouple feeds signals to a meter whose pointer indicates a value on a scale. This means that the observer must write down the meter readings as and when they happen. It would be much better if the procedure could be made automatic. A **chart recorder** is one way of collecting readings with no observer present. The electric signals cause a pen to draw a continuous graph on a strip of paper. More recently, **digital recorders** have become widely used. These present the readings as a number, and the numbers can be stored in a computer until needed. One virtue of electrical instruments is that they are very versatile.

Redox potential

A section in chapter 9 is devoted to oxidative phosphorylation. It is explained how absorption spectra can show the order in which the hydrogen and electron carriers function. Another approach to the same sequence of reactions is through measuring **redox potentials**.

'Redox' is short for 'reduction/oxidation'. A carrier is reduced each time it receives an electron; and the compound the electron came from is a reducing agent. The carrier is oxidised when it passes the electron on to the next compound in the sequence. For example, flavine adenine dinucleotide (FAD) passes electrons to one of the cytochromes. The reason it does this, and does not instead receive electrons from the cytochrome, is that FAD is a stronger reducing agent than the cytochrome. It has a greater tendency to get rid of the electron it is carrying.

Now this transfer of electrons from a stronger reducing agent to a weaker one is something like an electric current. When a potential difference causes a current, the electrons flow from a negative potential to a more positive potential. Similarly, in redox reactions a stronger reducing agent is said to have a more negative redox potential than the compound it passes electrons to. Moreover, redox potentials are measured in volts or fractions of volts, just like electric potentials.

To measure a compound's redox potential, a container is filled with a solution of the compound in both its oxidised and reduced forms. Also, an electrode is supplied connected to an electric circuit. The electrode is just a metal rod. However, one compound's redox potential cannot be measured on its own. It must always be compared with some other substance. Therefore, another container is prepared like the first one, but using the second substance. The electric circuit links the two containers, called **half-cells**.

In figure 63a, one half-cell contains a mixture of hydrogen ions and molecules. The redox potential of a hydrogen half-cell is arbitrarily given the value zero. It is the standard against which all other redox potentials are measured. Let us suppose that the other half-cell contains FAD and its reduced form $FADH_2$; the $FADH_2$ can be thought of as FAD with two extra electrons and also two hydrogen ions. Now, in each

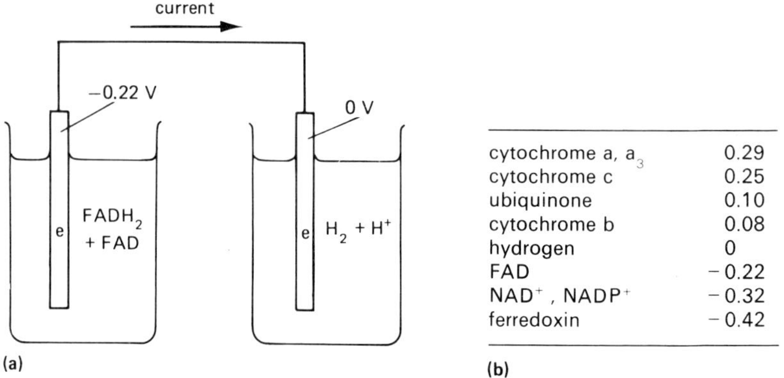

cytochrome a, a_3	0.29
cytochrome c	0.25
ubiquinone	0.10
cytochrome b	0.08
hydrogen	0
FAD	−0.22
NAD^+ , $NADP^+$	−0.32
ferredoxin	−0.42

(b)

Figure 63 Measuring redox potential. **(a)** Each half-cell contains a substance in its reduced and oxidised forms. The electrodes contain electrons (e) which are taken from solutes or given to them. An electric potential difference between the half-cells causes a current to flow. **(b)** Redox potentials of some biologically important compounds, measured in volts at pH 7 and 298 K (25 °C).

half-cell the reduced form (hydrogen molecules or $FADH_2$) has the opportunity to give its extra electrons to the electrodes. A molecule may bump into the electrode, which absorbs the electron, the molecule then remaining in the solution as its oxidised form. While the electrode can absorb electrons in this way, it can equally easily provide electrons. An oxidised molecule such as FAD may bump into the electrode and pick up an electron, thus becoming reduced. But the total number of electrons in the electric circuit must remain the same: if one electrode absorbs electrons, the other gives out an equal number.

Which way, then, do the electrons go? They can be absorbed by the left-hand electrode, flow as a current to the right-hand electrode, and there be given to the hydrogen ions. Alternatively, the current can flow the opposite way. Which happens depends on how strong the hydrogen and $FADH_2$ are as reducing agents. A stronger reducing agent gets rid of its electrons more powerfully than a weaker one. In this case $FADH_2$ is the stronger, so it gives electrons to the electrode, turning into FAD in the process. At the other electrode, hydrogen ions have to accept electrons and turn into hydrogen atoms and molecules. The current flows because an electrical potential difference is set up between the two half-cells, in this case 0.22 V measured under certain standard conditions. So the redox potential of $FAD/FADH_2$ is -0.22 V, and that of H^+/H_2 is of course 0 V. The minus sign means that $FADH_2$ is a stronger reducing agent than hydrogen.

Figure 63b lists some biologically important substances and their redox potentials. You may like to compare them with the chains of carriers used for phosphorylation in chloroplasts and mitochondria. However, redox potentials ought to be treated with some caution. Values given in books are measured under fixed conditions of temperature, pH, and the ratio of oxidised and reduced molecules in the half-cell. A change in some of these conditions can markedly alter the reducing power of a substance. In particular, if there were many more molecules of FAD than $FADH_2$, its reducing power would be less, and it might then be capable of accepting electrons from a cytochrome. This could happen in a living cell if some other reaction was producing large quantities of FAD, or using up $FADH_2$. So, although redox potentials help us understand oxidative phosphorylation, they are not a foolproof guide to which way electrons actually travel in living protoplasm.

Moving-coil ammeter

One of the most commonly used output transducers is the moving-coil **ammeter**. This measures current, but with simple modification it can also be used to measure voltage.

In some ways, the ammeter works like a compass needle. A compass needle is a small bar magnet, producing a magnetic field all around it. The earth is also a magnet, producing another magnetic field. These two fields interact with each other; in particular, the needle experiences a turning force which only stops when the two magnetic fields are lined up.

In chapter 8 it is pointed out that an electromagnet can also be used to produce a magnetic field. There it is used for focusing the electron beam in an electron microscope. An electromagnet consists essentially of a coil of wire, with many turns, carrying an electric current. In the ammeter, there are two electromagnets: one is fixed, like the earth, and the other is free to rotate like the compass needle (figure 64). Unlike the compass needle, though, the strength of the moving magnet can vary. It increases if a stronger current flows through it, so the ammeter is actually designed to show the strength of the moving magnet.

A compass can work satisfactorily with a needle of more or less any magnetic strength. It comes to rest pointing in the same direction. But this would not be much use in an ammeter; there needs to be some way of allowing the moving magnet to rotate further when it is stronger, that is, when there is a larger current flowing through it. The device used is usually

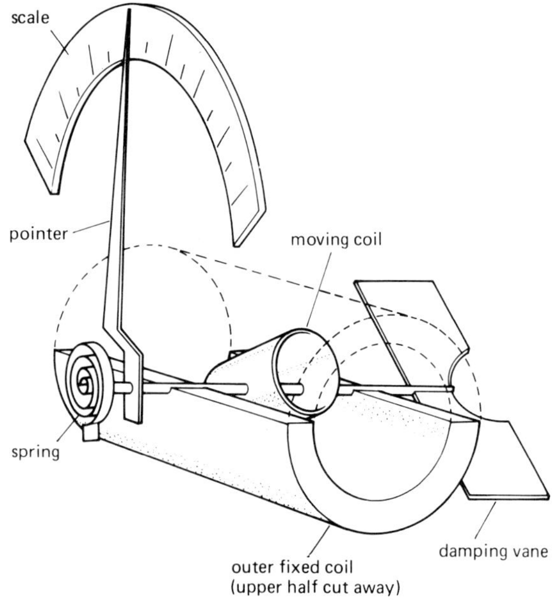

Figure 64 Moving-coil ammeter. The outer coil, scale and outer end of the spring are attached to the meter's frame. The inner end of the spring, the pointer, inner coil and damping vane are fixed to the rod, which is free to rotate.

some kind of spring. Then, if the meter is well designed, the rotation of the moving magnet and the pointer attached to it is proportional to the current in the moving coil. These meters can be very reliable and accurate. They have something like a one per cent error when the needle is deflected fully; this means, of course, that they are not really very sensitive.

Pros and cons of ammeters

Moving-coil types of meters are built into many instruments. As they are so widely used, it is worth giving some thought to the problems attached to them.

The poor sensitivity is not so serious. If better sensitivity is needed, a more sophisticated version of the meter can be used. But it is worth remembering that when an error of one per cent is quoted, it usually refers to a reading made at the top end of the scale. If a reading is made at the lower end of the scale, the same error will be a much larger percentage. Suppose the reading is only 0.05 amp (A) when the maximum value on the scale is 1 A, then the error becomes twenty per cent. This is not very sensitive by anyone's standards!

Even so, this kind of meter is useful for measuring something like the temperature of air in a forest canopy. The variable being measured does not fluctuate much, so it is not important if the observer has to wait about 1 s for the meter needle to settle down to a steady reading. Unimportant variations in the air temperature from minute to minute are probably greater than the instrument's error anyway.

But will an ordinary ammeter serve to record a nerve impulse passing along a neurone? The important difference is that the neurone event is short-lived and changes rapidly. In just two milliseconds (ms), the ordinary resting potential of − 60 mV suddenly rises to about + 60 mV, drops to about − 80 mV, and finally rises to the ordinary resting potential again. Imagine for a moment we are a team of electrical engineers faced with the problem of designing a suitable moving-coil meter; how can we make the meter respond quickly enough to record the three stages?

The first approach would probably be to cut down the mass of all the moving parts. This would reduce their inertia, allowing the meter to respond quickly to a change in voltage or current. But the problem then would be that the needle would flash across the scale to the correct value, only to overshoot. A meter needle that overshoots will wobble to and fro around the true reading until hopefully it stops in the right place. But in the case of the nerve impulse, the + 60 mV we are trying to measure has by this stage changed back to the resting potential. The meter fails miserably to record the events in the neurone.

Faced with overshooting, we might design some form of damping.

This is usually achieved electronically, and cuts down the wild fluctuations of the needle. Another simple solution is to attach a damping vane (figure 64); this slows down the needle's movement by creating air resistance. However, damping inevitably slows down the rate at which the meter responds to a sudden event. So, we are almost back where we started.

Fortunately for the neurophysiologist, the electrical engineers' brilliant solution is to introduce a completely different kind of output transducer. Having no inertia, no fluctuation, and giving an instant response, this transducer makes any effort to improve the moving-coil ammeter seem futile.

Cathode-ray oscilloscope

The instant response of an **oscilloscope** is due to the negligible mass of the 'moving parts'. This is because the pointer is replaced by a beam of electrons, and there are no other moving parts at all. Therefore, the problems of fluctuation and damping do not exist. These advantages mean that an oscilloscope can be used as a superior kind of voltmeter or ammeter. It can also perform other tasks of great value to the biologist. In particular it can record accurately what happens during complicated and very fast events, such as the passing of a nerve impulse.

An oscilloscope's screen, like that of a television, is the flattened end of a cathode-ray tube (figure 65). The cathode-ray tube is mentioned in chapter 8 as the output transducer of a scanning electron microscope. An image, or other kind of display, is projected onto the screen by means of a narrow beam of electrons. This beam is produced by an electron gun at the far end of the tube, rather like the one used in electron microscopes. Cathode-ray tubes are so called because the electrons, being given off by a heated cathode at high negative potential, were originally called cathode rays.

In the scanning electron microscope, the picture on the screen is built up like that of a television; the electron beam crosses the screen many times, filling it with horizontal lines. In the oscilloscope, however, the arrangement for moving the beam is different. The electrons first pass between two horizontal plates. If these plates are maintained at the same electric potential, they have no effect on the beam, but if the upper plate is more positive than the lower, the electron beam is attracted upwards slightly. This makes it strike the screen higher up. The greater the potential difference between the plates, the larger is the vertical deflection of the beam.

For a similar reason, the potential difference between the next pair of plates controls the horizontal deflection. If the right-hand plate is about

100 V more positively charged than the left-hand one, the electron beam is pulled across to the right so that it strikes the edge of the screen. These high voltages are needed on the plates because the electrons are propelled at very great speed from the electron gun.

Since the vertical and horizontal deflections can be varied independently, the electron beam can be directed to any chosen point in the grid marked on the screen. It is just like plotting a point on a graph. The point can be seen because on the inside of the glass is a layer of '**phosphor**'. This is a material which gives out light when bombarded by electrons. Various 'phosphors' are available. Some are not really phosphorescent, since most of the light is given off immediately from any particles struck by the electrons. Technically speaking, this is **fluorescence**. Where it is useful to keep the image for a short while after the electrons stop arriving, a truly **phosphorescent** material is used. This is something that continues to give off light even though electrons are no longer striking it. A mixture of 'phosphors' can be even more useful. One commonly used mixture gives an immediate blue light, and fades into a green afterglow. It may still be visible up to 10 s after the electron bombardment has ceased.

Although the horizontal deflection can express any variable, just like the horizontal axis of a graph, it is often used for time. Then it is known as the **time base**. In figure 65, the screen is displaying the change

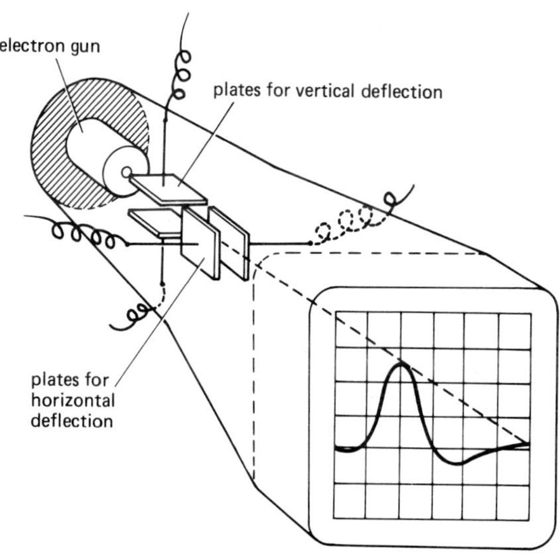

Figure 65 Cathode-ray oscilloscope. An electron beam, shown as an interrupted line, passes from the electron gun, between the two sets of deflecting plates, to strike the screen.

in potential difference across a neurone plasma membrane as a nerve impulse passes. As described earlier, the resting potential of – 60 mV gives way to a positive value of about + 60 mV, followed by a sharp fall and a slight rise. All this takes place in 2 ms, so the time base of the oscilloscope must be adjusted to last just over 2 ms. This is arranged by making the potential difference between the vertical plates steadily increase over that period of time. The electron beam therefore crosses the screen from left to right at an even speed.

However, the observer may not know exactly when the nerve impulse is going to pass his electrodes (figure 62). This is no problem for the oscilloscope. The time base can be triggered by the nerve impulse itself. In other words, the disturbance of the resting potential in the neurone can be made to start the horizontal deflection of the electron beam. In this way, the graph plotted out by the electron beam is placed in the middle of the screen.

As mentioned earlier (p. 126), the tiny change in voltage at the neurone plasma membrane needs to be amplified. From bottom to top, the graph on the screen represents about 140 mV, but the change in voltage at the horizontal plates must be more like 100 V. A convenient feature of oscilloscopes is that the amount of amplification can be varied. In this case it is simply adjusted to make the expected graph just the right size to fill the screen.

By now it should be clear that the oscilloscope is tremendously versatile. Just one or two more applications will be mentioned, to increase our admiration even further. Suppose, for example, that you are studying nerve impulses, and you need each one to be displayed as large as figure 65. This may be because you want to measure the time an impulse takes to pass and the change in potential. But suppose you also want to measure the time elapsing between successive impulses in a volley. If this is tens or hundreds of milliseconds, obviously a whole series of impulses cannot be fitted onto the screen at once; the time base is only 2 to 3 ms. One ingenious way around the difficulty is to switch off the time base altogether. If this is done, the spot formed by the beam merely moves up and down, forming a vertical line. Now a camera is mounted on the oscilloscope to photograph the screen. But instead of keeping the film still while a photograph is taken, the camera shutter is opened and the film is wound through the camera at a steady speed. If the speed is chosen carefully, a nerve impulse appears on the developed film just as shown on the screen in figure 65. The movement of the film is a perfect substitute for the time base of the oscilloscope. Winding the film can be kept up for much longer than 2 ms, of course, so, effectively, the oscilloscope display can be made indefinitely wider without losing any information.

So far it has been assumed that the time base, when it is switched on, only sweeps across the screen once. Often, though, it is made to sweep across the screen repeatedly, just as in a television or scanning electron microscope. This can be done at a very high rate, many thousand times a second. Each time the voltage between the vertical plates reaches its maximum, it suddenly drops to zero. That makes the electron beam, and therefore the spot, return almost instantaneously to the left-hand side.

There are many applications of this. Take, for instance, the time-of-flight mass spectrometer described in chapter 4 (figure 15). This is an input transducer, and the oscilloscope can be a suitable output transducer for it. Suppose that we need to find out how much $^{16}O^{16}O$ and $^{18}O^{18}O$ there is in the oxygen produced during a photosynthesis experiment. One pulse of gas results in two lots of ions arriving at the cathode, and these are detected as two very small bursts of current. Now this would be impossible to measure using a moving-coil meter. It is much easier to record the graph of current against time on the oscilloscope screen, just as is done for the nerve impulse. If 50 pulses of gas are fed into the mass spectrometer each second, the time base of the oscilloscope is made to operate also 50 times per second. Each time, the same graph of current against time is produced on the screen. Because it is happening so fast, the display looks static, and it can be kept on the screen without fading as long as one needs to measure it.

12 Pressures

How much pressure is required to keep a bird up in the air? How much is needed for us to breathe in or out, or to drive blood around our bodies? What is the filtration pressure in the kidney glomeruli? How do fragile-looking plants manage to break their way through concrete using nothing but osmotic pressure? Clearly there are many intriguing questions concerning pressure.

In chapter 3 (p. 33) it was described how Stephen Hales measured a horse's blood pressure. It is worth considering why Hales' method would not be used today. For one thing, it presumably placed the animal under considerable physiological stress: Hales' horse died during the experiment. Even if the measurement was accurate, therefore, it might not have been typical of a normal animal. But in any case the measurement was probably inaccurate because a lot of blood was withdrawn from the horse. It would be surprising if this did not significantly lower the pressure. You can probably think of other drawbacks in Hales' method.

Manometry

However, the idea of allowing pressure to push liquid up a tube is very sensible. This is still used today in instruments called **manometers**. A glass tube is bent into the shape of a letter 'U' and half-filled with some liquid. For measuring large pressures, a dense liquid such as mercury is used. If both ends of the tube are open to the air, the liquid stays at the same level in the two halves; the diameters of the tubes make no difference. But if the air is drawn out of one half, which is then sealed, the mercury rises up that side until its level is 760 mm above the level on the other side. The manometer is now a mercury **barometer**. For convenience we say that the strength of the atmospheric pressure is 760 mm Hg.

This way of expressing pressures is convenient and established by long usage. It is still used in medicine. Unfortunately, several other units are now also employed. In meteorology, 760 mm Hg or one atmosphere is almost exactly one bar. One bar is equal to 10^5 Pa or pascal. The pascal is the S.I. unit, and is defined as $1\ N\ m^{-2}$ (newton per square metre). These units are listed because they are all used by biologists. Although rather antiquated, the atmosphere unit is quite useful. For instance, the pressures inside some plant cells reach 30 atmospheres. This emphasises

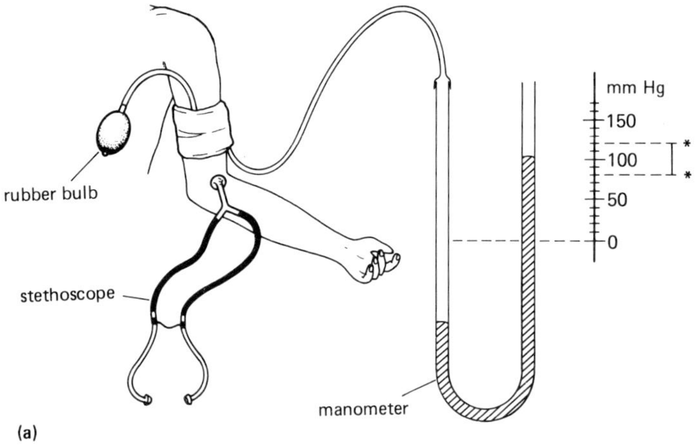

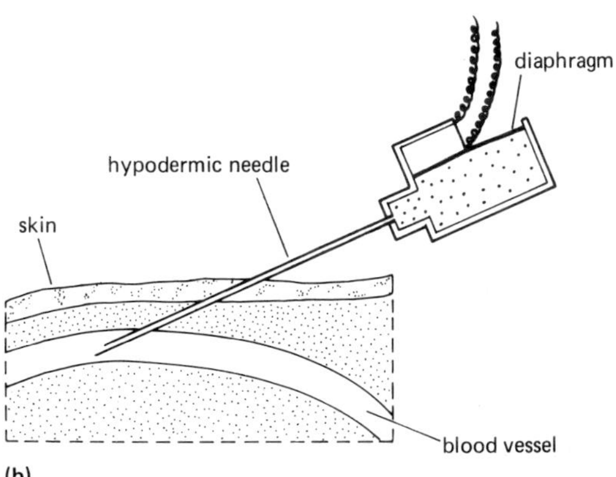

Figure 66 Blood pressure. **(a)** Auscultatory method of measuring systolic and diastolic pressure in an artery. Air pressure in the rubber bag around the arm is raised using the bulb until it stops blood flowing in the artery. The pressure then falls, and the points are noted where the pulsing noise of blood flow begins and ends (marked with asterisks). The scale is the height difference between the two mercury levels. **(b)** Pressure transducer. Air in a rigid box takes up the same pressure as the tissue around the needle tip. The greater the pressure, the more the diaphragm is pushed out; so the lower is the electrical resistance of the wire stretched from the diaphragm to the transducer frame.

the enormity of these pressures, and helps to answer one of the questions at the start of this chapter.

There are several ways of measuring blood pressure. The main method used in the doctor's surgery actually measures two pressures. One is the systolic pressure in the major arteries; that is, the maximum pressure produced each time the heart contracts. The other is the diastolic or minimum pressure.

An inflatable rubber bag is wrapped around the upper arm, and pumped full of air (figure 66a). A mercury manometer or some other meter is used to measure the air pressure. At first, the pressure is pumped up well above 120 mm Hg. This is higher than the normal systolic pressure, so the arteries leading to the lower arm are squeezed shut. No blood gets through, and so no noises are heard through a stethoscope placed on the inside of the elbow. Now the bag is slowly deflated. As the pressure falls, so does the mercury level. When it reaches about 120 mm Hg, the air pressure just equals the systolic pressure. As soon as it falls lower, once in every cardiac cycle a little blood passes through the artery under the bag. This is because the blood pressure momentarily exceeds the pressure in the bag. The pulsing of the blood flow can be heard through the stethoscope. One can also see the mercury in the manometer bobbing up and down. The pulsing noise continues until the air pressure equals the blood's diastolic pressure of about 80 mm Hg. Below that, the air pressure is not great enough to shut the arteries at any point during the cardiac cycle; so the blood flows through continuously.

In a healthy young adult, the average pressure in the aorta is close to 100 mm Hg. This pressure decreases continuously along the route the blood takes. At the beginning of the capillaries it is down to about 30 mm Hg, and drops to 10 mm Hg at the end of the capillaries. The quantities given here are the amounts by which the blood pressure exceeds that in the right atrium where it is zero. For measuring these low pressures in the veins, water could be used in the manometer instead of mercury. Water would rise 13.6 times higher in the tube than would mercury, and so the measurement would be more sensitive.

Manometers are fine for measuring fairly steady pressures, but what about fast-changing pressures, such as those during the cardiac cycle itself? The mercury in a manometer has so much inertia or mass that it cannot respond fast enough. To obtain a quick response, manometers have to be put aside and an electric transducer used instead. There are various designs, and some consist of a box mounted on the end of a hypodermic needle (figure 66b). The box is rigid except for one side which is a flexible diaphragm. When the needle is inserted into a blood vessel, the air in the box takes up the same pressure as the blood. If the pressure increases, the diaphragm bulges more. One method of measuring the diaphragm's movement is to stretch a wire from the diaphragm to a rigid support. If the diaphragm bulges the wire gets shorter, and this reduces its resistance. A potential difference across the ends of the wire will therefore make a larger current flow than when the wire is stretched longer and thinner. Some pressure transducers can respond in less than 0.01 s.

Osmotic pressure

In a typical human capillary, the blood's hydrostatic pressure is about 18 mm Hg above atmospheric. Measurements have also been made in the interstitial fluid just outside the capillaries. Surprisingly, the pressure here turns out to be lower than atmospheric, by about 6 mm Hg. In shorthand, this is expressed by calling it −6 mm Hg. The pressure difference across the capillary wall is therefore something like 24 mm Hg.

The pores in the capillary wall are on average 8 nm in diameter. They are quite large enough to let most of the plasma pass through freely. It might be expected, therefore, that the pressure would force water and solutes into the surrounding interstitial fluid. But another surprise is that this does not happen, at least not midway between the two ends of the capillary system. Why is this?

For a satisfactory answer we need to consider **osmotic pressures**. If a semipermeable membrane separates a solution from water, water molecules move through the membrane into the solution. This is osmosis. The reason appears to be that more water molecules hit the membrane on the water side than on the solution side; thus their ability to escape from the water is greater. But the water molecules in the solution can be made to escape faster through the membrane if the pressure of the solution is raised. When it is raised so much that water is escaping from the solution at the same rate as it is from the pure water at ordinary pressure, the extra pressure on the solution is called its osmotic pressure. Unlike ordinary hydrostatic pressure, osmotic pressure is a measure of a solution's ability to *absorb* water molecules.

Most of the solutes in the blood pass through the capillary walls quite easily, so the capillary wall is not a semipermeable membrane like the one just mentioned. It does, however, stop the protein molecules leaving the blood. Now the osmotic pressure of a fluid caused solely by dissolved proteins is called its **colloid osmotic pressure**. In the interstitial fluid there are few proteins, and in humans the colloid osmotic pressure is only 4 mm Hg. In the plasma, though, it is as high as 28 mm Hg, because there are many proteins (figure 67). The difference is 24 mm Hg, and this represents a pressure forcing water and small solute molecules into the blood. It is osmosis due to dissolved proteins.

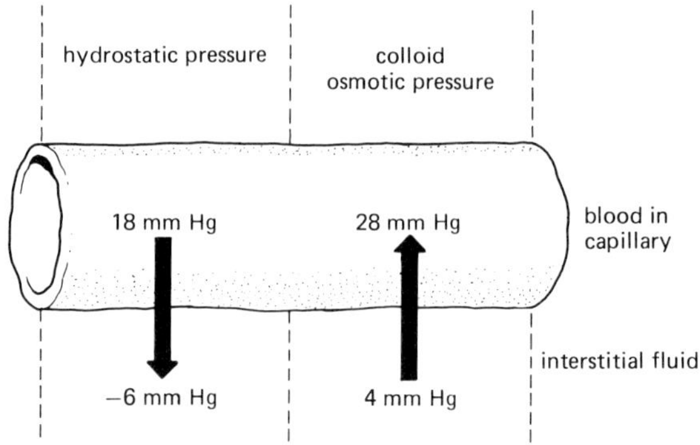

Figure 67 Pressures in a human capillary and outside it. The arrows indicate the flow of liquid expected from the hydrostatic pressures alone, and from the colloid osmotic pressures alone.

So, half-way along a capillary system, the colloid osmotic pressure just balances the ordinary hydrostatic pressure. The result is that plasma is neither lost nor gained by the capillaries through their walls. At the arterial end of the capillaries, however, the hydrostatic pressure of the blood is higher, because it is nearer the heart. This forces fluid out of the capillaries. Then at the venous end of the capillaries, the hydrostatic pressure has dropped lower than 18 mm Hg, so, at this point, the colloid osmotic pressure can force water and solutes back into the blood vessels.

Measuring osmotic pressures

From this example it should be clear that it is not enough to measure merely hydrostatic pressures. The behaviour of blood does not make sense without measurements of osmotic pressures too. How, then, do we measure osmotic pressures? They *could* be measured in the way described above, by exerting a large pressure on the solution in question. But this is not easy; for one thing, the membrane will often burst or start leaking. It is also a very slow method, and needs a fairly large amount of sample. Fortunately, easier methods are available.

Biologists often need methods which can be applied to minute quantities of material. This is because only tiny amounts are available. For instance, micropipettes enable the glomerular filtrate from a single Bowman's capsule to be withdrawn, or the sap from a single algal cell. The fluid from a sieve tube can similarly be obtained by allowing an aphid to insert its proboscis until the cell is penetrated. Then, by cutting the proboscis from the aphid, the phloem fluid is collected as it exudes from the exposed end of the proboscis.

Given such a minute sample, the biologist can resort to the following method. A solution's osmotic pressure depends on the number of particles other than water in a unit volume, and this in turn determines the freezing point. The more solute particles, the lower is the freezing point compared with that of the pure solvent. Therefore, a careful determination of this temperature and a quick calculation will yield the osmotic pressure.

An alternative method is available. It looks simple, but is in fact surprisingly sensitive. A range of standard solutions is made up. We shall call these A, B, C, D, and E. They all have different osmotic pressures. Then five glass capillary tubes are taken, and each filled with alternating drops of a standard solution and the sample (figure 68). The tubes are placed side by side on a slide before being stuck down and sealed with a blob of wax at both ends. Then the length of each sample is measured under a microscope.

After a few hours the sample drops are measured again. If by chance solution C has the same osmotic pressure as the sample, the latter will not have changed length. This is because the air bubbles between the drops act like semipermeable membranes; only water can pass from one drop to another by evaporating and condensing at the other end of the bubble. If solutions A and B have higher osmotic pressures than the sample, they will have absorbed water from the sample drop, so that will have decreased in length. Similarly, if D and E have lower osmotic pressures they will have given water to the sample. At first solutions A to E can be chosen to cover a wide range of osmotic pressures; this gives a rough estimate of the sample's. Then another series of standard solutions can

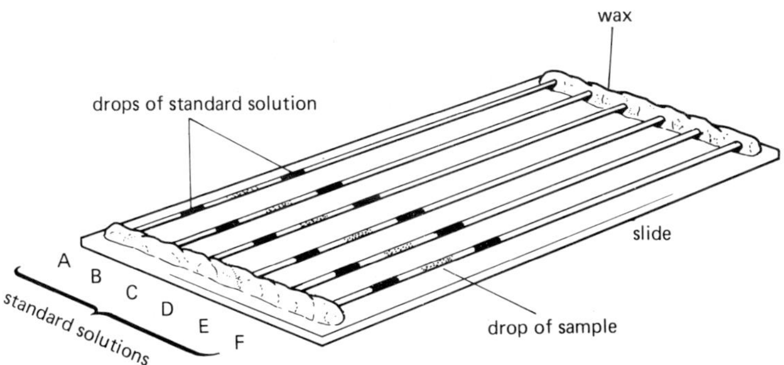

Figure 68 Measuring osmotic pressure by the vapour-pressure method. In each capillary a drop of the sample is surrounded by drops of a standard solution of known osmotic pressure. The sample drop which does not change length has the same osmotic pressure as the standard solution in the same capillary.

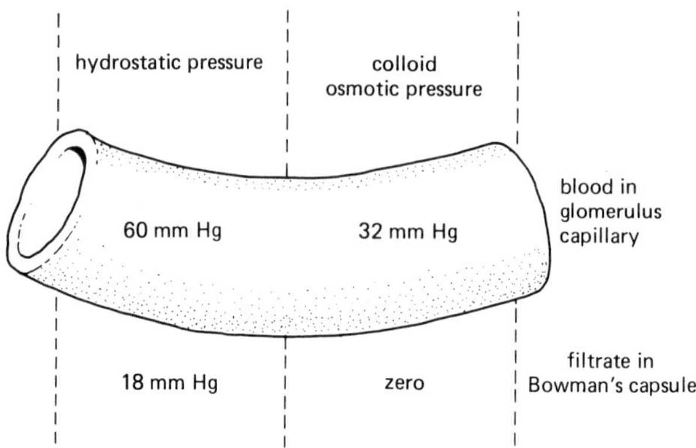

Figure 69 Pressures in the human glomerulus and Bowman's capsule. How high is the filtration pressure?

be used, covering a much narrower range. Working in stages like this, a good sensitivity can be achieved.

As another example of the application of these methods, you might like to examine figure 69. It shows the hydrostatic and colloid osmotic pressures in the human glomerulus and Bowman's capsule. Taking into account both kinds of pressure, how many mm Hg are responsible for the glomerular filtration? This is the filtration pressure. You may be surprised at how small it is. If the hydrostatic pressure in the glomeruli decreased by this amount, the kidneys would stop working.

Warburg manometer

So far this chapter has considered some important pressures in biological materials. Manometers can, however, be used in a rather different way. Since it is so easy to measure pressures which do not change too fast, manometers have been adapted to study a whole range of other biological processes.

The best-known apparatus is the Warburg manometer. It is named after the German biochemist Otto Warburg. Some of his most important achievements are mentioned in chapter 9 (see p. 111 and figure 55): for instance, he discovered cytochrome oxidase and was the first to measure its action spectrum. To measure the yeast's oxygen consumption he used a manometer, and the work gained him a Nobel Prize in 1931. We have a special reason to remember Warburg, since he admired the English and regularly took his holidays in England. Although born in 1883, he died only in 1970.

To one side of the Warburg manometer (figure 70) is attached a flask in which the reaction takes place. The other side is open to the air. Therefore, any difference in the two fluid levels shows the pressure of the gas in the reaction flask. Normally, the pressure itself is of little interest: but it can easily be used to show changes in the amount of gas. At constant temperature and volume, the pressure is proportional to the quantity of gas. For instance, a ten per cent increase in pressure means there are ten per cent more molecules.

In principle, then, a Warburg manometer can be used to follow any reaction causing a change in quantity of gas. A number of biological examples spring to mind: photosynthesis, aerobic respiration, and fermentation. Where only one gas is involved, as in fermentation, there are no problems. However, if two gases are changing at the same time, they both contribute their effects. In this case their effects may be distinguished by using two manometers. Imagine we are studying the aerobic respiration

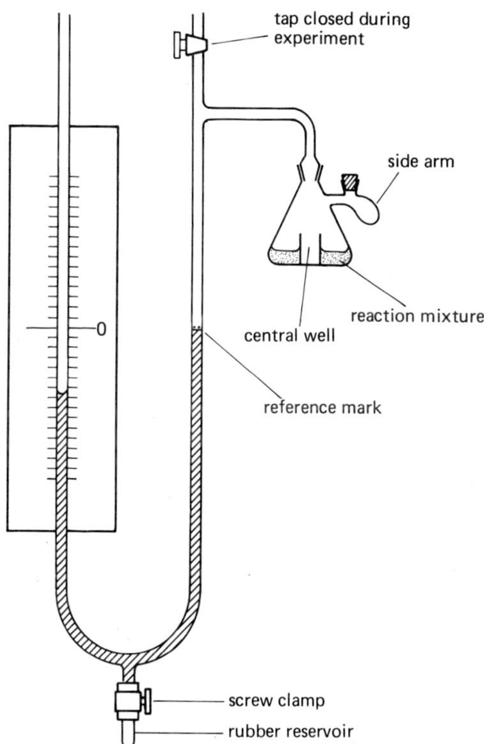

Figure 70 Warburg manometer. The reaction being studied occurs in the bottom of the flask. At regular intervals the pressure of the gas is measured by bringing the right-hand fluid level to the reference point marked on the glass; this is done by adjusting the volume of liquid in the rubber reservoir, using the screw clamp.

of bacteria by trying an enzyme inhibitor such as cyanide. One manometer could be measuring the change in both oxygen and carbon dioxide. Meanwhile, the second one contains a strong alkali in the central well of the flask. This absorbs any carbon dioxide given off, so in this manometer, the pressure changes are due solely to the oxygen being used up. The difference between the pressure changes in the two manometers tells us the rate at which carbon dioxide is given off. If in another experiment we need to absorb oxygen, during photosynthesis for instance, chromium (II) chloride can be placed in the central well instead of the alkali. Further variation in experimental design is possible using the side arm. This can contain some substance such as cyanide which is to be tipped into the reaction mixture part of the way through the experiment.

These are examples of biological processes using up or producing a gas. However, the Warburg manometer is more versatile than this. For instance, we may wish to study lactic acid fermentation, in which glucose is converted into lactic acid. To measure the rate, a hydrogencarbonate buffer is added to the reaction mixture. Each molecule of lactic acid produced reacts with a hydrogencarbonate ion (HCO_3^-) to release a molecule of carbon dioxide. The rate of the reaction can therefore be followed with the manometer.

Simple though the Warburg manometer is in principle, it is subject to various errors which must be allowed for. For instance, the act of removing a gas may affect the reaction. An example of this is the rate of photosynthesis, which is dependent on oxygen concentration. Another consideration applying to any gas is that not all of it is in the sealed air space of the flask and manometer; some is dissolved in the reaction mixture. This can be allowed for by calculation, assuming that the dissolved gas is in equilibrium with the gas in the air space. To ensure there is equilibrium, the flask is mechanically shaken throughout the experiment. Another reason for the shaking is to maintain a constant temperature, and to ensure this the flask is also immersed in a water-bath.

Two things which can seriously affect the manometer readings are the atmospheric pressure in the laboratory, and temperature. The pressure is important because it is pressing down on the open side of the manometer. On the other hand, any rise in temperature will increase the gas pressure in the flask, simply because it inevitably acts like a thermometer filled with gas instead of mercury. To get around these problems of pressure and temperature, a **thermobarometer** is usually kept in the water-bath along with the manometers. This is really a control; it is made up just like the experimental manometers except that it has no reactions taking place in it. Any changes it detects in the surrounding temperature and pressure are subtracted from the readings of the experimental manometers. This corrects for the unwanted effects.

Postscript

This book has been about instruments and how to make observations. Why is it good to know about these things? There are several reasons.

First, knowing where biological facts come from helps with learning biology. For instance, a student who understands why proteins can be separated by electrophoresis will thereby know more about proteins. A similar comment can be made about every other tool and technique described in the book.

A second reason should appeal particularly to students who want to become scientists, either as a career or a hobby. Some biologists have good ideas but are not very clever at making observations, let alone designing instruments with which to make them, so the biologist who enjoys instruments and can even build totally new ones has a tremendous advantage. Consider radioisotope dating, X-ray diffraction, pollen analysis, the electron microscope, and many other techniques: when first developed, each of these opened the way to a wealth of exciting discoveries and valuable new theories based on them. If you are fortunate enough to invent or develop a new instrument or method in biology, you have a good chance of making some important advances.

There is another kind of biologist who does not like theoretical work very much, but is talented in the practical aspects. For such people there is a good career to be found, not only in research institutions but also in hospitals and industry. The technician who is expert with instruments can be an indispensable member of a research team.

Yet another reason is that good science cannot flourish without careful criticism of new theories, and even of generally accepted opinions. Invalid arguments, poor evidence, lack of objectivity in observations – these are all blemishes which need to be recognised and avoided if biology is to develop successfully. A knowledge of how observations are made, and particularly of the errors to which instruments are subject, can help produce a high standard of scientific work.

I hope this message emerged when you read the earlier parts of this book. In contrast, the rest of the book is crammed full of dogmatic and authoritative statements! My only excuse is that this was necessary for clarity and brevity. Still, I hope you have read it all with a critical eye – *and* that you found no mistakes.

References

References chapters 1–3

W. I. B. Beveridge (1968) *The Art of Scientific Investigation.* (Mercury Books.) A short practical guide to biological research, with many examples.

D. C. Carter, M. S. Gosden, A. Orton, G. T. Wain, C. Wood-Robinson (1981) *Mathematics in Biology*, 'Selected Topics in Biology' series. (Thomas Nelson.)

J. Goodfield (1981) *An Imagined World.* (Hutchinson.) This gives a blow-by-blow account of how one investigation was carried out. Long and detailed.

G. H. Harper (1967) Polymorphism in a Bardsey slug. *Bardsey Observatory Report*, number 14, 61–63.

O. V. S. Heath (1970) *Investigation by Experiment*, 'Studies in Biology' series. (Arnold.) Short and readable; includes statistical aspects.

F. R. Harden Jones, G. P. Arnold, M. Greer Walker, P. Scholes (1979) Selective tidal stream transport and the migration of plaice (*Pleuronectes platessa* L.) in the southern North Sea. *Journal du Conseil international pour l'Exploration de la Mer*, **38**, 331–337.

D. W. Tucker (1959) A new solution to the Atlantic Eel problem. *Nature*, **183**, 495–501.

M. Greer Walker, F. R. Harden Jones, G. P. Arnold (1978) The movements of plaice *Pleuronectes platessa* tracked in the open sea. *Journal du Conseil international pour l'Exploration de la Mer*, **38**, 58–86.

J. D. Watson (1981) *The Double Helix.* (Weidenfeld and Nicholson.) Another account of a famous piece of research; short and readable.

References chapters 4–12

Each book covers a wide variety of instruments and methods with the exception usually of microscopy.

R. McNeill Alexander (1975) *The Chordates.* (Cambridge University Press.)

R. McNeill Alexander (1979) *The Invertebrates.* (Cambridge University Press.) These two undergraduate textbooks are easy to read and include a variety of measuring techniques.

H. Hillman (1972) *Certainty and Uncertainty in Biochemical Techniques.* (Surrey University Press.) Fairly advanced; a knowledge of techniques is assumed.

R. H. Kay (1966) *Experimental Biology.* (Chapman and Hall.) Advanced.

D. W. Newman (editor) (1964) *Instrumental Methods of Experimental Biology.* (Collier-MacMillan.) Advanced, and very useful for reference.

R. W. Van Norman (1963) *Experimental Biology.* (Prentice Hall International.) A readable introduction.

J. A. Ramsay (1965) *The Experimental Basis of Modern Biology.* (Cambridge University Press.) Only slightly more advanced than this book; well worth reading.

B. L. Williams, K. Wilson (editors) (1981) *A Biologist's Guide to Principles and Techniques of Practical Biochemistry.* (Arnold.) Fairly advanced, but good for reference.

References chapters 4–12

Each concentrates on one method of research.

A. L. E. Barron (1965) *Using the Microscope.* (Chapman and Hall.)

Sir W. L. Bragg (1975) *The Development of X-ray Analysis.* (Bell.) Well illustrated, but fairly advanced.

H. R. Wilson (1966) *Diffraction of X-rays by Proteins, Nucleic Acids and Viruses.* (Arnold.) Elementary and well illustrated.

W. Burrells (1977) *Microscope Technique.* (Fountain Press.)

H. Carter (1957) *An Introduction to the Cathode Ray Oscilloscope.* (Philips, Eindhoven.) Elementary and readable.

J. M. Chapman, G. Ayrey (1981) *The Use of Radioactive Isotopes in the Life Sciences.* (Allen and Unwin.)

C. F. A. Culling (1974) *Modern Microscopy.* (Butterworths.)

A. M. Glauert (1979) Recent advances of high voltage electron microscopy. *Journal of Microscopy*, **117**, 93–101.

G. W. Gray (1951) Electrophoresis. *Scientific American*, **185**, 45–53.

A. V. Grimstone (1977) *The Electron Microscope in Biology*, 'Studies in Biology' series. (Arnold.) Short and readable.

I. Hickman (1981) *Oscilloscopes: How to Use Them, How They Work.* (Newnes Technical Books.) Up-to-date and fairly advanced.

R. S. Sharpe (1979) Projection microradiography, *Journal of Microscopy*, **117**, 123–143. Describes the X-ray microscope.

W. H. Stein, S. Moore (1951) Chromatography. *Scientific American*, **184**, 35–41.

B. S. Weakley (1981) *A Beginner's Handbook in Biological Transmission Electron Microscopy.* (Churchill Livingstone.) More advanced than Grimstone; good for reading and reference.

C. A. Williams (1960) Immunoelectrophoresis. *Scientific American*, **202**, 130–140.

References chapters 4–12

These give further details of studies mentioned in the book.

H. Hillman, P. Sartory (1980) A re-examination of the fine structure of the living cell and its implications for biology teaching. *School Science Review*, **62**, 241–252.

D. Keilin (1966) *The History of Cell Respiration and Cytochrome.* (Cambridge University Press.) Of historical interest; very detailed.

M. Meselson, F. W. Stahl (1958) The replication of DNA in *Escherichia coli. Proceedings of the National Academy of Science*, **44**, 671–682. Reprinted in G. L. Zubay (editor), (1973) *Papers in Biochemical Genetics.* (Holt Rinehart and Winston, New York.)

R. H. Michell, J. B. Finean, R. Coleman (1982) Widely accepted modern views of cell structure are fundamentally correct: a reply to Hillman and Sartory. *School Science Review*, **63**, 434–441. This paper and the one by Hillman and Sartory summarise the debate about artefacts. It can also be followed on a taped discussion: H. Hillman, R. Gliddon (1982) *A New Look at the Structure of Living Cells.* (Audio Learning.)

J. D. Watson, edited by G. S. Stent (1981) *The Double Helix.* (Weidenfeld and Nicholson.) Besides Watson's book, this edition contains much additional material, including the important original papers announcing the double-helix theory.

Index